THIS BLESSED EARTH

ALSO BY TED GENOWAYS

The Chain (2014)

Walt Whitman and the Civil War (2009)

Anna, Washing (2007)

Bullroarer: A Sequence (2001)

THIS
BLESSED EARTH

A Year in the Life of an
American Family Farm

TED GENOWAYS

W. W. NORTON & COMPANY
INDEPENDENT PUBLISHERS SINCE 1923
NEW YORK | LONDON

Photographs by Mary Anne Andrei

This Blessed Earth is a work of nonfiction. The Hammonds, Galloways, and Harringtons are a real family in York and Hamilton counties, Nebraska. All quotations derive from recorded conversations between July 2014 and October 2016. Whenever possible, conversations have been depicted just as they occurred between the harvest of 2014 and the harvest of 2015; however, occasional additions from follow-up interviews have been included for clarity and detail. In some cases, names of individuals have been changed.

"Fear in a Handful of Dust" © 2015, Ted Genoways, first published in *The New Republic*. "Corn Wars" © 2015, Ted Genoways, first published in *The New Republic*. "The Great Abandonment" © 2016, Ted Genoways, first published in *The New Republic*. "Irrigation Nation" © 2016, Ted Genoways, first published in *Pacific Standard*. "This Blessed Earth" © 2017, Ted Genoways, first published in *Harper's*.

For information about permission to reproduce selections from this book, write to Permissions, W. W. Norton & Company, Inc., 500 Fifth Avenue, New York, NY 10110

For information about special discounts for bulk purchases, please contact W. W. Norton Special Sales at specialsales@wwnorton.com or 800-233-4830

Manufacturing by LSC Communications, Harrisonburg
Book design by Brooke Koven
Production manager: Beth Steidle

Library of Congress Cataloging-in-Publication Data

Names: Genoways, Ted, author.
Title: This blessed earth : a year in the life of an American family farm / Ted Genoways.
Description: First edition. | New York : W.W. Norton & Company, [2017] | Includes bibliographical references.
Identifiers: LCCN 2017025062 | ISBN 9780393292572 (hardcover)
Subjects: LCSH: Family farms—Nebraska—Case studies.
Classification: LCC HD1476.U62 N25 2017 | DDC 630.9782/345—dc23
LC record available at https://lccn.loc.gov/2017025062

W. W. Norton & Company, Inc.
500 Fifth Avenue, New York, N.Y. 10110
www.wwnorton.com

W. W. Norton & Company Ltd.
15 Carlisle Street, London W1D 3BS

1 2 3 4 5 6 7 8 9 0

Now farming became industry . . .

—JOHN STEINBECK,
The Grapes of Wrath

Contents

READYING THE BIN

September 2014

The last yellow hues of evening were fading to blue. A bright circle of sun blazing through an uncapped vent in the roof had climbed the corrugated side of the grain bin and smoldered out like an ember. Now the only remaining light, there in the dark interior, was the flame of Kyle Galloway's cutting torch—first a flickering orange tongue, then as he adjusted the hissing valve, a bright cerulean cone of fire. Sparks showered and bounced across the concrete slab as he cut another steel plank down to size.

Still in his twenties, Kyle was broad-chested, with dark, buzz-cut hair and a gentle round face and glasses that gave him a kindly, almost child-like air. But he moved through his work with quiet authority, inching across the width of the plank until the unwanted end clanked on the concrete. Kyle paused a beat to let the fresh cut cool, then he laid the plank into place, the next piece in a floor now nearly half done, and stomped it again and again with his heel. Each boot fall roared loud as thunder, shaking the whole structure, until the section snapped tight. Kyle stepped up, bouncing to test its soundness. Then, satisfied, he leaned through the open portal door.

Outside, in the gathering dark, Dave Tyson, the farm's hired man, handed over the next section of floor. Dave wore a ball cap and round glasses of his own. Easily twenty years Kyle's senior, he seemed to be making a conscious show of keeping pace. The sleeves of his flannel shirt were torn off, and the last of the summer heat formed a film of sweat on his pale biceps, the muscles tensing and flexing as he hoisted the flooring. Salvaged from an older bin on a distant part of the acreage, these planks had been laid up in a defunct hog shed for years. But nothing ever goes wholly to waste on the farm. Old tractors and field trucks, hay rakes and a faded harvester all stood tucked away in leaning barns or grown over by tall grass at the edge of a shelterbelt, dormant but awaiting purpose. For this flooring, it was finally here.

At the end of June 2014, crop counters at the U.S. Department of Agriculture had released estimates that the summer's mild weather and increased rainfall would mean a much-higher-than-expected yield. It's the kind of prediction that sounds like good news, but for farmers it meant an upcoming glut of grain. One report projected that the national soybean yield would be 3.3 million acres higher than expected at planting—not just cutting into profits but actually outstripping domestic demand. The outlook for corn was little better. Though acres planted to feed corn were down, the stockpiled reserves from the record yield of 2013 remained at an all-time high. On news of such unprecedented surpluses, corn and soybean futures nosedived among speculators at the Chicago Board of Trade. After several years of riding high on $8 corn, farmers watched prices fall to $3.50 per bushel. Soybeans, which had topped $17, dropped below $10.

Then in August, in the weeks just before harvest, a series of downpours moved across the middle of Nebraska, dumping two and three inches of rain at a time, pushing back the start date for bringing in the beans. While Kyle and Dave waited anxiously for fields to dry enough that a harvester wouldn't mire in the mud, projections for the national yield kept going up and up. Every day that the crops stayed out, unharvested and undelivered to the co-op, the per-bushel prices went lower. For weeks in nearby towns, at the firehouse in Hordville and the bar in Polk, even at the Iron Skillet at the truck stop south of York, the talk had turned to the Jews in Chicago making bank on the backs of farmers. Kyle didn't have time for that. He rolled his eyes and shook his head at old-timers who were drinking coffee and bullshitting instead of doing what they could to cut their losses.

Kyle didn't own this acreage, just south of the Platte River in east-central Nebraska, but his girlfriend, Meghan Hammond, was part of the sixth generation of her family to live and farm in this area—and Meghan and Kyle were starting to talk about a wedding in the next year. Once that happened, Meghan's father, Rick, figured it would be time to give them a stronger hand in the operation. Whenever people asked Rick what he raised on his farm, he always responded, "Corn and cattle and kids." By the end of the year, if all went as planned, Meghan and Kyle wouldn't be kids anymore.

But handing over control of the operation—"succession,"

as it's known—is one of the hardest times in the life of any family farm. The older generation always struggles to let go of the reins, to trust their kids to carry forward a long and fragile tradition. The younger generation bridles against the meddling and second-guessing and feels the double weight of scrutiny and doubt in every misstep. Many farms don't officially change hands until the older generation dies, leaving the rising generation beholden to their parents until they're well into adulthood and often into their fifties or sixties. It breeds resentment and too often divides generations or turns brothers and sisters against each other. It is so stressful and legally complex that there are psychiatrists and attorneys alike who build whole practices around helping families navigate the process. But on the Hammond farm, it was even more complicated than that.

The collapse of commodities prices was forcing the family to wrestle with how best to run the farm in the future. Years ago, Rick had tried raising organic corn and direct marketing his grass-fed black Angus cattle, but those efforts had largely failed. Now, he was not only planting crops resistant to a range of herbicides for commercial sale but actually growing genetically modified seed corn for DuPont Pioneer to sell to other farmers. He said he couldn't say no to the guaranteed rates that Pioneer pays those who raise their seed stocks. Over the years, that increased income had allowed Rick to expand his operation, but now he'd come to rely on it. Without the premium from Pioneer, a down market, like the one his family was facing in the coming year, would place the whole farm at risk. And even Pioneer was not impervious to the vagaries of the Board of Trade. With per-bushel prices so low, the company was worried that there would be less demand for corn seed in 2015. So Rick had been informed that Pioneer intended to reduce the number of acres they were planting— and, with it, reduce payments to the farmers who planted for them. The company just didn't know where it would cut yet or by how much. "The seed corn is about half of what we raise," he told me. "If they gave us all that we try to shoot for, it's more like three-fourths."

Without that money coming in, Rick knew that he couldn't afford to buy new machinery for the year, couldn't afford to replace his center-pivot irrigation system. He needed to have a full year of paying down those loans, not incurring new debts. He needed a year in which everything went right— crops came in early and healthy, prices rebounded, equipment held out. And if he was forced to take out more loans, he needed interest rates to stay low. "We're trying to tighten our belt and get rid of machine debt, pivot debt, and then get it down to land debt," he told me.

And that penny-pinching would be all the more important as the family faced the still-uncertain fate of the proposed Keystone XL. That pipeline project, intended to carry heavy crude pumped from the Alberta tar sands in Canada nearly 2,000 miles to Port Arthur and Houston on the Texas Gulf Coast, was slated to cut right through major portions of the family cropland. Meghan and Rick vehemently opposed the project, arguing that it would not only further slash their crop production but that it also posed an existential threat to the Ogallala Aquifer, the underground reservoir of freshwater that they depend on to irrigate crops and water their cattle.

Rick even agreed to let the antipipeline nonprofit Bold Nebraska build a solar-powered barn, the so-called Energy Barn, on land he rented from his in-laws. The publicity had brought in an unexpected amount of national press, culminating in Willie Nelson and Neil Young joining forces with Bold Nebraska to mount the fund-raising Harvest the Hope concert in the cleared cornfield of Art Tanderup, a farmer near Neligh, Nebraska, whose land was also threatened by the pipeline. But unpublicized amid all the hoopla was the reality that the barn, built in the middle of a productive piece of farmland, had cost Rick still more precious acres of ground, where he had planned to plant soybeans that year.

Worse still was a letter that arrived soon after from one of Rick's neighbors. He had rented hundreds of acres to Rick for many years, but now, in protest, was pulling out of that long-standing agreement—a move that was likely to cost the Hammonds between $150,000 and $200,000 of gross income for the coming year. The cancellation letter made no bones about the reasons: "I am sorry to inform you that your outspoken endorsement of defeating the Keystone Pipeline is not in our common interest. While it is certainly your right to support whatever causes you wish, we have chosen not to subsidize those expenditures."

With these financial hurdles ahead, Rick had decided to take some calculated risks. He planted almost entirely soybeans for the fields not already set aside to raise seed corn for Pioneer, because the bean prices seemed steadier, and he went with short-season beans, hoping to get the harvest in ahead of the inevitable market decline. Some giant grain companies, like Archer Daniels Midland, will even pay out a premium early in the harvest. "I planted really short-season beans, planning on getting it to ADM," Rick told me, "because there might be a dollar bonus above your elevator if you can get it in before the rush." He also made the decision to wait on marketing his beans, rather than locking in spring prices. Most years, he would have seen spring futures as good enough to hedge against fluctuations, but this year he'd decided to take a chance. "I'm not nearly as forward contracted this year as I usually am," he said.

That crop, waiting to be harvested and sold by the truck-load at whatever the market would eventually bear, was where the family was either going to make money or see a whole year's labor turn into a break-even proposition—or worse. With the weather conspiring to increase the national yield and push prices lower by the day, while leaving Rick's fields too soggy to get out and harvest those short-season beans, he was losing money with every passing hour.

So Rick had decided to lay a concrete footing and move in an old grain bin. He would put up as much of the soybean harvest as possible—and, with luck, wait out the down market. When he and Kyle brought the bin in and put it in place with a crane, he explained that even this stubbornness was its own form of risk. Grain in the bin is given to rot—and, of course, there's always the chance that prices will continue to fall. But you control what you can. Don't increase your risk by investing in a new bin, but make sure the old one is tight and well-ventilated. Make sure the floor is raised so an industrial blower can force through enough air to keep the grain dry. And after everything that's already happened, be realistic.

"You have some hardheaded old-timers that say, 'By God, beans were at seventeen dollars two years ago, and I'm going to leave my beans in the bin until it's seventeen dollars again.'" Rick shook his head. "You want to say: 'At your age? Good luck with that.' It's not going to happen." So, in your heart of hearts (though you'd never admit it, not even to your neighbors), you pray for rootworm in Chile, for hail on the open plains of Illinois, a late-season catastrophe unforeseen by the USDA. You hope for what the traders call a "market recovery," knowing that what that really means is that, in order for you to succeed, someone else as careful as you, through no fault of his own, has to fail.

In the last light of the day, Kyle stretched the tape measure from one wall of the bin to the other, marked the spot on the next floor piece with a nub of blue chalk. Then he was back at the torch. He struck the igniter, and the nozzle burst again into flame, sparks skittering and bounding across the concrete as he made another cut. By the time he finally stopped work for the day and walked back to the house, it was full dark. Meghan was already in the kitchen, back from nightly chores—feeding the goats and horses, checking on cattle, closing up the laying hens until morning. She had four chicken breasts cooking in the fry pan. Kyle kicked off his boots in the mudroom and padded quietly around the kitchen. Worry was heavy in the air. With harvest now two weeks late, Meghan was on edge too. The smallest provocation could turn her fiery eyes, usually wide with emphasis, to a cold stare. She was the first to admit that growing up on the farm with a twin sister and two other siblings had made her quick-tempered and sharp-tongued, but those years had armed her with a winning sense of humor. She could punch you in the arm hard enough to convey honest frustration and simultaneously, somehow, forgiveness.

This tension was different. It wasn't nervy but nervous. Meghan and Kyle both knew that the later in the season the harvest begins, the deeper in the year you go toward winter— and the chance of yield-loss due to ice or snow. If you suffer crop losses when prices are down, there's no chance of catching a market rebound. Then you become the sucker whose losses faraway farmers cheer. You take what the insurance adjustor gives you and try to make it through the next year. Flipping the chicken breasts with a fork, Meghan let out an exasperated moan, then launched a sardonic recitation of the wisdom she'd heard a thousand times. "Two things with farming," she said, putting on her dad's voice, "you can't control the weather, and you can't control the markets."

Meghan loves to joke about Rick's penchant for teachable moments, but she knew that if her dad was going to step back and eventually retire, then she and Kyle not only had to work hard. They had learn to make the smart decisions that give them the tiny edge that is often the difference between the kind of success that allows you to add equipment and acres and the kind of loss that eventually leads to a farm sale. With so much riding on the next year, Meghan said she would feel a lot better if they could just get out in the fields.

"You're putting inputs all year, with money and hard work and lots of effort," she said. "And one bad storm could wipe it all out." She moved from the stovetop to the sink, filling water glasses and staring out the window, as if she could see the brittle rows of ready soybeans and the browning cornstalks beyond the edge of their yard—though it was now so dark that the glass was a mirror, reflecting her face by the kitchen light.

"It's your payday," Meghan said, "the one time of year that we're actually making money, instead of shelling out." She took a deep breath. "It's a field of dollar bills out there."

Kyle cracked a wry smile. "And they could just blow away."

PART ONE

BRINGING IN THE BEANS

October 2014

The bright-green hulk of our John Deere harvester crept across the field of soybeans. It was late in the day, early October, the sun low. A cloud of hulls and chaff spewed from the back of the combine, then swirled up around us and blazed in the glow. Sealed in the dustless quiet of the cab, Rick Hammond steadied the wheel with one hand and punched coordinates into the touchscreen onboard computer with the other. With each new piece of data entered, the system let out a high-pitched beep or a low, lazy bloop. The reel of the harvester head spun steadily below like the paddlewheel of a river steamer, standing up stalks for the toothed blades cutting a dozen rows at a go. The feed auger corkscrewed the cut plants into the mouth of the combine, where a throbbing set of threshers splintered the dry pods, collecting the oily seeds inside and sending them spiraling up to the grain tank behind Rick's chair. Harvested beans ticked against the back window like a light summer hail. The only other sounds over the churning reel were the *Pong*-like beeps from the computer.

"Okay," Rick said at last, marking final coordinates, "we're on autosteer now." And, to show me, he took his hands off the wheel. "I thought I'd be able to retire before I had to get autosteer," he sighed. Rick is in his sixties, tan and weatherworn, but still plenty fit. Most days, he works sunup to sundown, especially during harvest. It wasn't the long hours he said were getting to be too much but trying to keep up with the technology. "The reason for all this is inputs," he explained. The combine continued along, following the contours of the planting lines automatically recorded months earlier by the Global Positioning System. As we moved, our progress was charted on the touchscreen in varying colors to show where each row or part of a row was above or below the target for bushels-per-acre for this field. All of that data is recorded and stored to plan for next year, helping farmers decide how to adjust the density of their seed populations, where to apply fertilizer, how much to water, where to put inputs and where to save. "During the era of high prices, yahoos that should have been broke were thinking, 'There's nothing to this.' Now we're back down to cost of production—or less. So everything matters. Every place you can cut a corner or save, you do it."

Even being out here was a tough call. Earlier in the day, with the sky clear and a light breeze blowing, Rick had asked Dave to check out the condition of this field. "Dave poked his nose in here just to get the combine set," Rick said. He found that the soybeans were dry enough to harvest, but the ground, after weeks and weeks of rain, was so wet that bringing in all the equipment they'd need to work the field—the harvester, the tractor and grain cart, the semi to haul the beans away— was going to compact the soggy soil and make it too dense to plant next year. After some thought, Rick decided to make the sacrifice. "It was scheduled to be seed corn next year," he told me. But Pioneer had already informed its seed growers that the company would be further reducing fields, from 30,000 acres in York County in 2013 down to 15,000 in 2015. "We were scheduled for six fields," Rick said, but he expected that Pioneer would reduce that number by at least one field before •

planting came around in the spring. "So I told them, 'You know what? If you're going to short us a field, this one would be the one.'"

The decision might seem impulsive—why not wait just a few more days?—but this urgency arises from a particular oddity of soybean biology. "Beans are weird," Rick told me. "With corn," he said, pointing to the stalks in the neighboring quarter, "it's growing degree units, the heat." Then he pointed straight ahead to the horizon, the sun orange and sinking low. "With beans," he said, "it's sunlight." So, unlike most crops, which begin to mature when the weather turns cold and the air dries, soybeans are short-day plants— meaning, simply, that their final stages of maturity are triggered by waning hours of sun.

This poses two major problems when you run into an unusually wet fall. First, the days get shorter no matter what. Weather-dependent crops like corn slow their maturation on rainy days. You can get lucky—catch a warm spell or even a single sunny day—and get out in the field and back on track. Not so with soybeans. While you're waiting for the clouds to clear, the beans are going past maturity. Second, soybeans have to be delivered to the grain elevator at a precisely defined moisture rate: 13 percent. "We're at twelve-point-nine right now," Rick showed me on the touchscreen, "but when we get down there where the ground is a little wetter, it'll go back up." Above 14 percent, he said, the grain elevator not only charges you for drying but also docks you for the estimated shrinkage. "And you would think, 'Well, one percent of moisture would be one percent of shrink,'" Rick told me, "but no. They dock you *one and a half* percent." At 15 percent, they're apt to reject the whole load.

So to make your best profit on soybeans, you need a sunny day (but not *too* sunny) with a dry breeze (but not *too* dry), and you need that day to fall exactly when the plant has received the precise number of hours—yes, *hours*—of sunlight from the moment you planted it months earlier. To make hitting such a tight window even remotely possible, seed companies, like Rick's supplier DuPont Pioneer, have hybridized soybeans for nearly a century—and genetically modified them in recent decades—according to bands of latitude, called "maturity groupings." They number these photo-regions from zero in the northern growing zones of Canada up to seven in the light-drenched flatlands of Florida. If you're lucky enough to have ground in the middle of a maturity grouping, then you can buy a single variety. But Nebraska is almost exactly divided between groups two and three, the line bisecting the state into north and south. Which means that farmers here, especially in central regions like York County, plant both varieties to spread out maturity, allowing them to harvest one group or another first, depending on the weather, and still get a consistent yield. Some daring farmers like Rick will formulate a guess as to what the weather holds for the growing season and plant more of one group or another, hoping for higher yields and higher returns.

In 2014, after several years of drought, Rick bet on another dry year—and planted incredibly short-season beans. While most of his neighbors were planting 3.5 maturity groupings, Rick had planted 2.4. And he was dead on, right up until the rains started in August. If he could have gotten out early, ahead of farmers in other parts of the country, and caught the market at its peak, he was positioned to make up for all the other setbacks going into the harvest. But if you guessed right on the growing season, as Rick did, and then got an extremely wet fall, you could end up with hundreds of acres of mature soybeans and fields too wet to run the combine. With each storm that rolls up on the horizon, you move from making a hefty profit to incurring a crippling debt.

Rick knew prices would hold for a time, on the chance of an unforeseen, late-season catastrophe, a hedge against an ice storm freezing crops or a line of thunderheads dropping hail that could send prices skyrocketing. But as grain started pouring into the elevators, prices on futures were sure to slump. With the market already in free fall, some farmers had decided not to wait: they went out as soon as the mucky furrows would allow and used propane or electric dryers to deal with the high-moisture grain. But this was yet another expense, and if the moisture levels were too high, the cost of drying could cancel any profit. Other farmers, like Rick, had held out as long as possible, but eventually everyone had to make hard choices.

"Every day that the beans sit out there," Rick said, "you're under risk of a big, smashing storm. And then, every day that we wait, the days get shorter. And beans get harder and harder to get out, because they just soak up moisture like crazy. And then, the more that happens—when they're dry, wet, dry, wet, after they're mature—they're prone to shatter. They'll split wide open in a big wind." So when the rains finally let up, he decided it was time to go, soggy fields or no. It wasn't worth risking the return on this field this year just for the promise of above-market prices on seed corn next year. He had Dave set up the harvester, and he started across the field to judge for himself if the beans were ready.

Now halfway to the end of the swath, Rick checked the

level of the grain tank. He needed to empty it out before he turned the combine and aimed back toward the barn and cluster of outbuildings on the eastern edge of the property. He radioed over to Kyle on the CB, "Can I dump on you?" Kyle pulled up alongside the combine with the tractor and grain cart, moving in perfect parallel. While the harvesting reel kept spinning and the combine inched across the field, the unloading auger arm started pouring out soybeans, sputtering and thrumming until the tank was empty and Kyle peeled off to unload into the trailer of the big rig. Rick's mind was already back on the moisture levels. At the end of the swath, he took a wide sweeping turn and set the harvester back on autosteer.

"This here is instant yield, instant moisture," Rick said, pointing to the screen, "and this is average yield, average moisture." As we moved, he could see in real time if he was on track to hit his production targets or falling short and whether the moisture of the entire load was within the acceptable range or inching high enough that he'd have to pay a penalty for drying. He watched the data rolling across the screen, giving the minute-to-minute condition of the crop, like watching the stock market ticker rise and fall. The colors of our current twelve rows shifted back and forth from green to yellow, from profit to loss. All the perils of modern farming seemed to crawl across the four-inch screen, teetering from making a living to accruing debt.

So far, everything looked fine. "They're about thirteen and a half," Rick said. "We'll get a little dock, but that's okay. A year like this, if you can get a couple of loads in, you just keep pecking away at it." So he continued across the field, until the sun was almost gone and the night cool made the beans too wet to cut. By then, Kyle was at the wheel of the harvester and had cleared the clogged cutting blades four times, trying to get just a few more minutes in. Finally, he came over the radio. "It's time," he said. "It's past damned time." And with that, everything was over for the night. Dave and Kyle parked the tractor and seed cart next to the outbuildings, but left the combine precisely where it sat. Dead in its tracks until the sun and the wind came up again in the morning.

But harvest, at last, was officially underway. Now, Rick just had to get his crops in as soon as possible. It was going to be a race.

THE OVERPRODUCTION of corn by the American agricultural industry has generated a lot of attention in recent years, as food activists and environmentalists have grown worried about the middle of the country turning into a vast monoculture. But that's only half the picture. In truth, the corn boom, to a remarkable extent, has been made possible by a co-equal boom in alternate-year planting of millions of acres of soybeans. In many ways, corn and soybeans seem made for each other. Soybeans are a natural nitrogen-fixer, replenishing the soil for nitrogen-hungry corn hybrids, and the plants share almost none of the same pests or diseases, preventing insects, molds, and bacteria from overtaking fields. But the soybean is more than an enabler of King Corn; it is, in fact, far and away the most successful crop introduced to the American farm in the last century.

In 1920, there were fewer than a million acres of soybeans planted in the whole of the United States, well behind traditional staple grains. But soybean production boomed during the 1930s and after—surpassing barley and rye production by 1940, cotton in the 1950s, oats in the 1960s, and wheat and hay in the 1970s. Today there are more than 85 million acres of soybean fields in America, almost exactly equal to the acreage of cornfields, with which they are rotated from year to year, and almost all of those acres are concentrated in the Midwest and on the Great Plains. How did the soybean, a legume native to East Asia and traditionally used primarily in foods from China, come to have such a place of prominence here? The rise of the soybean in the United States is attributable to, more than any other person, Henry Ford.

The American farm underwent a period of unmatched innovation in the early twentieth century. The arrival of the gas-powered tractor for plowing, the combine for harvesting, and affordable farm trucks for hauling grain to market made it possible for farmers to plant more and more acres and to manage those acres with fewer farmhands. But by the mid-twenties, the market was glutted with grain, depressing prices and endangering the very family farms that the technological revolution had promised to empower. Farmers began calling for research and development to turn toward finding new uses for existing agricultural supply rather than continuing to search for ways to increase yields.

In January 1927, Wheeler McMillen, associate editor of the popular magazine *Farm & Fireside*, published a watershed article entitled "Wanted: Machines to Eat up Our Crop Surplus." He wrote that he had been receiving panicked letters from farmers, lamenting that more grain was being produced than people could possibly eat. Seeing as how "the human stomach isn't elastic," McMillen suggested that chemical compounds in plants might be converted into industrial products. He even advocated for government backing for such research. "There is no wrong in channeling some federal funds into farmers' pockets," he wrote, considering that the American farmer "by cheap food has subsidized the growth of cities."

Among McMillen's most interested readers was the owner of the Ford Motor Company. Ford, after all, manufactured much of the equipment that had contributed to booming yields—and if the company was to maintain that market share, then it had to find a way to sustain its customers. To Henry Ford's mind, it also made perfect sense to subsidize research into the uses of farm products, because he was already growing concerned about dwindling petroleum supplies. Numerous auto parts were made from petroleum-based plastics, and, of course, all of Ford's engines ran on diesel and gasoline refined from petroleum crude. "If we want the farmer to be our customer," he said, "we must find a way to be his customer." In early 1928, he met with McMillen to discuss this new field of research—what was eventually dubbed "chemurgy"—and came away even more convinced that the key driver should be private industry, not the government.

Soon after, Ford authorized dramatically expanding the agricultural laboratory in Greenfield Village at Ford's headquarters in Dearborn, Michigan. He appointed Robert Boyer, a self-taught but eager young chemist, to manage what came to be known simply as "The Chemical Plant" and to oversee its staff of a dozen or so young men from the Henry Ford Trade School. Under Ford's direct supervision, the lab technicians experimented with a staggering range of vegetables, fruits, grasses, legumes, tubers, and roots. Ford would meet with Boyer to hear new ideas—which plants might contain high levels of cellulose for plastics, which might contain sugars that could be converted to ethanol to make biofuels. The next morning, Boyer would find a truckload of fresh produce waiting at the front door of the lab for testing.

After the stock market crash in October 1929, the financial crisis for farmers deepened. Ford publicly advocated for continuing full production of all crops and again urged the government to resist stepping in. "The farmer and the chemist will solve farm relief, not the politician," he told the *New York Times*. Nevertheless, the USDA sent emissaries around the world in search of new crops that could be planted on American farms—crops intended for industrial use, not food. On one such trip to China, William J. Morse gathered more than 10,000 soybean varieties for American researchers to study. Learning of this, Ford urged Boyer to look at the soybean. His staff found that soy had unexpected levels of usable oils and yielded high-protein soy meal after the oil was extracted. It appeared, in short, that the soybean could be used to produce industrial lubricants and that the byproduct could be turned into plastics.

The results were so encouraging that, in December 1931, Ford approved $1 million in new research funding and instructed Boyer to halt work on all other plant life. The following spring, Ford had 300 varieties of soybeans planted on 8,000 acres on his farmland in rural Michigan. The next year, he expanded to 12,000 acres—making him the single largest soybean grower in the United States. That same year he announced that he would buy any soybeans delivered to the Dearborn plant. To encourage production, he made 400 Fordson tractors available for free use to Michigan farmers and offered gas and diesel at a penny per gallon—less than a quarter of what it cost at the pump. Farmers put more than 35,000 acres of land into growing soybeans, and Ford bought their entire output as promised. It was a daring move but also good business. Between 1931 and 1933, with few farmers buying trucks or tractors, the Ford Motor Company lost nearly $120 million. Henry Ford was convinced that turning those numbers around would require restoring the buying power of the American farmer.

To that end, Ford put the full muscle of his publicity machine behind promoting soybeans—and chemurgic applications for other crops as well. He hosted the national convention of the American Soybean Association. He helped organize and again played host to the Farm Chemurgic Council, which was promoting Agrol, an ethanol-blend alternative to gasoline, over the loud objections of the petroleum industry. And Ford gave a series of interviews, telling one reporter that he envisioned a time when much of an automobile "could be made from by-products of agriculture." He even announced a plan to decentralize Ford production by opening a constellation of factories in rural areas to manufacture plastic parts made from local soybeans. "He wants to make as many parts as he can in small factories, whose workmen will live nearby, each tilling his little plot of soil," reported *Fortune*. While talking to one reporter, Ford had an assistant bring to his desk a steering wheel, a distributor box, and a grab bag of small electrical parts. He bragged that he used soybean oil in his paint and to lubricate his casting molds. Soybean meal was reacted with formaldehyde to produce a thermoplastic resin, which was used to make gearshift knobs, window frames, distributor caps, and horn buttons. "There is a bushel of soya beans in every Ford car," *Fortune* declared.

"He is as much interested in the soya bean as he is in the V-8." Ford even switched the company commissary over to baked goods made with soy flour and ice cream made with soy milk.

To expand and expedite processing, Ford soon began experiments using hexane solvent for extraction of soybean oil. In 1933, he sponsored the Ford Exposition of Progress in New York City, where he debuted a glass model of an extractor that used hexane—and called on American companies to build full-scale versions. The federal government, the State of Illinois, and the Illinois Farm Bureau all contributed funds to take up the challenge. Within months, Archer Daniels Midland (ADM), a milling company which had been supplying linseed oil to Ford for its paint until Henry Ford ordered it replaced with soybean oil, purchased from Germany a 150-ton-per-day Hildebrandt hexane extractor. When it began operation on Blackhawk Street in Chicago in March 1934, it became America's first continuous solvent extractor—and the most modern soybean processing plant in the world.

At the same time, Ford unveiled a home version of the extractor, which he hoped would be used on a small scale by farmers to extract oil and sell it directly to the company. The enormous Ford exhibit at the Chicago World's Fair in 1934 included a diorama of a working soybean farm and a modern "Industrialized Barn" with one of the small extractors inside. "I see the time soon coming when the farmer will not only raise raw materials for industry but will do the initial processing on his farm," Ford said. "He will stand on both his feet—one foot on the soil for his livelihood, and the other foot in industry for the cash he needs. He will have a double security. Agriculture suffers from a lack of market for its product, industry suffers from a lack of employment for its surplus men. Bringing them together heals the ailments of both. That is my conviction and that is what I am working for."

That summer, Ford's promotions got an unexpected boost. The first of several years of severe drought, what soon proved to be the worst dry time in American history, engulfed 75 percent of the country. An apocalyptic dust storm carried 350 million tons of topsoil from the Great Plains all the way to the East Coast. Corn and wheat withered in the fields. Amid rampant crop failures, farmers harvested 23 million bushels of soybeans, a better yield than most crops and far better than its oil-producing competitors like linseed and canola.

In the spring of 1935, farmers planted soybeans in record numbers. In preparation for the coming harvest, Ford spent $5 million to construct his own soybean mill with solvent extraction units at the flagship River Rouge plant in Dearborn—and boasted that he had jumpstarted demand for soybeans nationwide. That year roughly 70 million bushels of soybeans would be harvested. Selling for 50 cents per bushel in the absolute depths of the Great Depression, soybeans were hailed as a godsend. They now generated more income for farmers than either barley or rye, and soybean trading had become so active and central to industry that the Chicago Board of Trade started offering soybean futures for the first time. By the end of the 1930s, the soybean harvest was approaching 100 million bushels per year. *Time* magazine declared Henry Ford "a bean's best friend."

But no sooner had industry begun to move away from petroleum than the world's largest oil reserve was discovered in Saudi Arabia in 1938. The barrel price of crude oil fell precipitously, and the demand for petroleum substitutes waned.

Ford continued to extol the virtues of soybean products, but the industrial applications for soybeans were now seen as impractical. But during the years when grain crops were failing, animal feeders had discovered that livestock fed on an oil-rich soybean diet bulked up quickly. With demand from Ford and other automakers no longer elevating prices, milling companies, hoping to capitalize on their existing supply and marketing structures, were eager to get into the soybean trade.

At the time, America's milling industry was still centered on St. Anthony Falls at the headwaters of the Mississippi River in Minneapolis, Minnesota. In the nineteenth century, the city had taken advantage of the strong current of the falls to turn millstones and its geographical location, upstream from nearly all of the United States, to bring its milled goods cheaply to market. General Mills and Pillsbury shared an undisputed hold over the flour market, so their smaller competitors were eager to find ways to capture similar shares of other grains. Chief among these new competitors were ADM, which, in addition to producing linseed oil for Ford's paint, also milled, processed, and stored vast quantities of crop-based products, and Cargill, which specialized in milling grains as feed for livestock producers. They eagerly built new soybean-processing plants, along the rivers and the Great Lakes from Minneapolis to Chicago, where the great stockyards were buying unprecedented quantities of soybean meal as animal feed.

In 1939, Shreve M. Archer of Archer Daniels Midland announced the construction of a new modern soybean plant and elevator in Decatur, Illinois. Again acquiring its extractor from Germany, this time ADM ordered a Hansa Muehle unit that employed the Paternoster design, carrying flaked beans in buckets on a moving chain. The unit, which had a daily capacity of 400 tons (roughly four times the size of any that had been used in Europe), made ADM the largest soybean processor in the world. In November 1939, the plant was ready for operation. The solvent unit occupied a five-story tower; storage capacity for 5 million bushels of soybeans was provided.

At the same time, Cargill launched an aggressive, even desperate, bid of its own into soybean processing. Company president John MacMillan, Jr., had inherited the family-run business just five years before and expanded so forcefully into corn processing during the corn-poor Dust Bowl years that the U.S. Commodity Exchange Authority and Chicago Board of Trade accused the company of trying to corner the market. When government investigators demonstrated a pattern of overbuying—acquiring as much as 80 percent of the total market share—in order to artificially drive up prices, the Chicago Board suspended Cargill and three of its officers from the trading floor. (The company would not rejoin the Chicago Board until 1962, after MacMillan's death.) The company established a downriver soybean-processing facility, which they dubbed Port Cargill. Not to be outdone, Pillsbury entered into a partnership with an independent miller who was preparing to open a new plant in Mankato.

By the harvest of 1939, soybean mills dotted the Mississippi River from one side of Minnesota to the other. Now the problem was supply. Before the end of the year, the American Farm Bureau Federation sponsored hundreds of events to encourage farmers to plant soybeans and instruct them on how to achieve top yields. For the first time, the Midwest was the center of soybean production in America, with

ADM leading all processers with six soybean plants in Decatur, Chicago, Minneapolis, Milwaukee, Toledo, and Buffalo. With harvesting and processing now at full capacity, the soybean producers were well positioned to take advantage when the United States was pulled into World War II.

When Hitler began his march across Europe, there were sudden scarcities of edible oils and fats—nearly 40 percent of which had previously been imported from Mediterranean countries. After Pearl Harbor, the U.S. government pushed agricultural production to achieve record output—and soybeans increased from 78 million bushels in 1943 to 193 million bushels in 1945. Yet Cargill and ADM had a terrible time fully capitalizing. While the production—and consumption—of soy products nearly tripled, President Roosevelt and Congress worked together to pass price-control legislation, creating the Office of Price Administration (OPA). Meat and poultry were hard to come by, so the OPA established fixed prices on animal feed.

After the war, however, when the government slowly started to roll back its strictures, feed prices rose sharply. As Ford had always feared, government involvement, instituting price controls and later subsidies, made farmers subject to the whims of federal farm policy—and the agribusiness interests that controlled that policy. But by then, Ford was too sick and old to keep up the fight. He turned the company over to his son, and barely a year later, Henry Ford died of a cerebral hemorrhage at his estate in Dearborn. Without his guiding vision, soybean production for industrial purposes waned sharply for several decades, but the grain millers saw a different opportunity.

Wartime rationing officially ended in 1946, and the American public was eager to celebrate. Like never before, they wanted French wine and a good steak. In the early months of 1947, national consumption of meat rose by more than 20 percent—a real increase of more than 1.7 billion pounds. Prices climbed so high that New York's mayor called for a congressional investigation. But the *New York Times* concluded that it wasn't price-fixing. "Americans are meat-hungry," the reporter explained. "There are just more persons consuming more meat than ever before in the history of the country." To ease high prices, the federal government began subsidizing corn and soybean production. Ever since, the prices of commodity grains have risen and fallen according to the demand of livestock feeders.

Ford may have failed to create a future of plastic cars and biofuels, but he successfully created a steady market for soybean producers—only now, their profitability was once again tied to the whims of the livestock market.

RICK WASN'T pleased. "Not good," he said, "not good."

It was past eleven in the morning, the sky still overcast and threatening. With the remaining fields near the home section too wet for work, Rick had loaded the combine onto a flatbed and brought it to a quarter section of dry land south of Interstate 80 where he had planted more very short-season soybeans. The wind seemed to be picking up enough to bring the moisture down to where he could harvest. He had asked Dave to run a test patch, just to be sure, only to find that an unusual number of the bean plants were "laid down"—that is, their stems were so flat to the ground that the spinning reel of the

John Deere couldn't prop up the stalks for the cutting blades. So many beans laid flat meant a measurable loss of yield, but it also suggested the possibility of something even worse.

"We had a couple of hailstorms come through here this summer," Rick said, kneeling in a furrow next to one of the flattened plants, "but our insurance adjustor didn't expect the loss to be near this bad." He plucked one of the plants, roots and all, out of the ground and flipped open his buck knife. Slicing longways, he split the stalk in half. "Aw, man, look at that," he said, showing the hollowed-out insides to Kyle and Dave, who had come down from the cab of the combine to see what was wrong.

With one glimpse of the stem, they each set to pulling up their own plants and cutting them open. After a few tries, Rick called out. "Here we go," he said. He called Meghan over too, then held up the sliced open stem for everyone to see: a tiny larva, a narrow white caterpillar, nestled right where the stalk meets the roots. "Stem borer," Rick said. "They warned us to be on the lookout for them this year, but we've never seen them here before." He turned and cursed, then flipped open his clamshell cell phone, and in a minute had the crop insurance adjustor on the line. "Yeah, we need you to come out and have a look," he said as he wandered back toward the road. Rick kicked clods of dirt as he went, sending up tiny brown clouds.

Meghan shook her head. "You spend all fucking year trying to get these crops the best yield," she said. "It's ready to be harvested. They're done, and when they're done, you got to get them out."

But now, this field would have to wait for the adjustor, which could be days or even a week, depending on how many claims were already waiting. When a call like this goes out, the adjustor comes with a tape measure and marks off 15 feet. Then he counts every standing plant and every dead plant to tabulate an estimate of the overall percentage of loss. Different levels of insurance pay out at a variable rate. You can insure 50 percent of loss for a relatively low premium, all the way up to 75 percent for the highest premium. If your damage is less than 25 percent, you just have to eat the loss.

You can also get supplemental coverage, but it only kicks in if there's at least 14 percent crop loss across your entire growing region. And even if you make your deductible and the supplemental insurance kicks in, you're not only sustaining at least a 14 percent loss but the payout on the difference—the loss above and beyond that threshold—is figured according to an adjusted price, calculated from the average of the spring and fall prices. So whether you were forward contracted or hoping to make a little extra by marketing at harvest, it's impossible to catch the price at its peak. And in a year like this, when Rick was hoping to bin as many beans as possible and wait out the market, there was no chance to catch a rebound. Even with the best insurance, the Hammonds were going to take a substantial hit on this field—and while they waited, there was the chance of still more loss.

"Hail, wind damage, snow," Kyle explained. "Right now those crops are really vulnerable."

"Eight inches of rain," Meghan said.

"Yeah, they just got big rains in southeast Nebraska," Kyle said. "It wiped out a lot of fields. They say a two-hundred-year event."

"It's also driving the price," Meghan said.

"That's right," Kyle agreed. "In the week since those rains, beans went up a dollar. That's probably how much yield has been lost. Because the beans are really sensitive. High wind will just knock the beans right out of the pod."

Rick clamped his phone closed and started back toward the field, moving quickly. "Tell Dave that we're not going to get it out, so that he can just bring the combine in," Rick said. "Let's get the head back on the trailer." He told Kyle that they would move everything over to a nearby neighbor's property. The neighbor, Seth, had gone in with them on the rental for the combine. Rick had thought that his short-season beans would be ready first, but that wasn't going to happen now. They could make the most of what remained of the day and then hope to find another field that was ready tomorrow. For now, Seth's half section was dry, so they might as well get everything over there.

Kyle and Dave went right to work unhooking the head of the harvester.

"We're at least two weeks behind," Rick said in my direction but almost to himself. "And it's just killing us."

FAILURE IS everywhere on the farm.

It hides in the long shadows cast by the barn at last light. It waits amid the dark stalks shifting in the fields before dawn. It's all around you, always lurking, always palpable but just out of view.

These farm failures can be dramatic and sudden: the death of a patriarch, his chest rolled over while repairing a tractor or his overalls twisted into a choking knot by the auger in the grain bin. With no will or simply no clear plan for succession, the farm can go under. Just as shattering can be a divorce or the death of a spouse. Legal bills and the stress of unsettled assets can turn a farmer's attention from keeping the books or watching the markets and somehow the business slips away. And, of course, failure can come from the fields, appearing like an undetected cancer—cobs stunted by drought, beans infested by cutworms.

More often, though, it creeps up, not one major disaster but a series of smaller missteps. Too little insurance in a drought year. Borrowing to build a hog barn just before the bottom drops out of the market. Buying more ground and more seed only to see commodities prices plummet. Often as not, failure comes from nothing more than a farmer overestimating his ability to service a loan.

Some would say that this is no accident. In the early 1970s, Secretary of Agriculture Earl Butz told farmers that a new day was dawning. They had to "get big or get out." He urged those who heeded his call to acquire as much land as they could afford and plant "fencerow to fencerow." Butz wanted American farmers to produce a steady oversupply of key grains, in order to undercut and control commodities markets to the disadvantage of our Cold War enemies. But the United States had considerably less landmass than either the Soviet Union or China. So the only way to outproduce them was to invest in every possible method of intensifying production—chemical fertilizers and pesticides, ending crop rotation in favor of monocultures, consolidating farmland and agricultural companies, and ultimately tinkering with the genetics of row crops to make them resistant to various herbicides but also tolerant of denser planting as a way of increasing per-acre yields.

Butz promised to use the emerging global economy to bolster prices. If we faced harvests in which supply outstripped demand, the United States would simply negotiate trade deals, using our economic might to artificially create a market. If we could flood global trade partners with massive quantities of grains, pushing sale prices below the cost of production, then Butz believed the world would have no choice but to buy from us. We could make our enemies—and even our friends—dependent on us to feed themselves.

In January 1972, Butz sold what amounted to our entire grain reserve to the Soviets. The following month, Nixon went to China and brokered a deal with Chairman Mao Zedong, allowing the importation of American corn and securing contracts for American companies to build thirteen of the world's largest ammonia-processing plants for producing fertilizer on Chinese soil. America's Communist foes regarded these moves as an agreement not to wage war through food. But Butz discussed the new agreements in terms of "agri-power" and stated it plainly: "Food is a weapon."

The trouble, as many critics saw it, was that producing at that volume meant relying on agribusiness, making us beholden to a handful of companies. Where family-owned farms had been numerous and diverse, the kind of small operations that had to produce a variety of high-quality products to insulate themselves against market fluctuations, multinational grain companies were centralized and large enough to capture extensive portions of critical commodities and turn a profit by doing nothing more than capitalizing on market uncertainties at strategic moments.

The most famous example came in 1973, when Cargill, which had only been allowed to return to the Chicago

Board of Trade a little over a decade earlier, now placed huge orders for U.S. soybeans via a Geneva-based affiliate, Tradax, making it appear that there was a pending soybean shortage. Prices skyrocketed—and eventually rose so high that President Nixon ordered a halt to all soybean exports to prevent domestic scarcity. With prices at record levels and foreign countries desperate to find new sources of soybeans, Cargill filled the canceled orders of other American companies with supply from their South American subsidiaries, commanding artificially inflated prices. When the U.S. embargo was lifted, Cargill canceled its Tradax purchases and by year's end saw their annual profits jump from less than $20 million to more than $150 million, despite an overall decrease in its volume of production that year. In short, knowing that food was now a matter of national security and that the government would intervene to protect it, Cargill (and soon other major agribusinesses) recognized that they could manipulate the market to produce greater profits, regardless of the success of each year's harvest—and regardless of the impact on American farmers.

A few years later, farmers saw firsthand the risks they now faced. At the end of December 1979, President Jimmy Carter enacted a grain embargo against the Soviet Union after Leonid Brezhnev ordered the invasion of Afghanistan. Ever since buying our national granary, the Soviets had been acquiring huge quantities of American grain—including 25 million tons of wheat and corn contracted that October for the coming year. Carter bristled at the idea that the Soviets thought they could stage an invasion when we controlled their food supply. "They think I don't have the guts to do anything," he told an aide. "You're going to be amazed at how tough I'm going to be." Carter announced the embargo and canceled the outstanding grain orders for some 17 million tons of corn and wheat. He issued a freeze on trading in grain futures for two days to allow the market to stabilize, but when it reopened, prices plummeted anyway—losing 20 percent of market value. Angry farmers marched on the USDA. One protest leader said, "We planted fencerow to fencerow like they wanted, and now this is what happens."

The Soviets found other grain suppliers to circumvent the embargo and lessen the impact of any future trade sanctions. As commodities prices fell, it became apparent: instead of making the world dependent on our grain supplies, we had grown reliant on their demand. American farmers continued to carry heavy overhead and service high-interest loans, preventing them from competing with foreign grain producers, even after the embargo was ended. American agricultural exports declined by more than 20 percent between 1981 and 1983, which, combined with decreased market prices, ended up meaning a nearly 40 percent reduction in farm income in a matter of just a few years.

Some would regard this as the ultimate failure of agriculture, but the emergent industry of consolidated agribusiness flourished during this time. Along with the old grain cartels, such as Cargill and ADM, former defense contractors, including Monsanto and DuPont, aggressively entered the food industry. Cargill grew from $2.2 billion in sales in 1971 to $28.5 billion in sales in 1981 by turning the profits from grain shortages into a diversified portfolio. They moved into value-added operations—milling grain for animal feed, making high-fructose corn syrup for Coke and Pepsi—then expanded and vertically integrated, acquiring feed elevators and meatpacking plants. By the mid-1980s, Cargill was not only the nation's top grain exporter but also the Number 1 egg producer, the Number 2 beef packer, and the Number 3 miller of corn and wheat.

Meanwhile, the market free fall touched off by the failed grain embargo created a catastrophic decline in consumer confidence that led to a national recession and, as inflation began to rise, eventually set the stage for the Farm Crisis. Prices on key grains stayed low, and overproduction soon drove them even lower. At the same time, the accrual of massive debt in the form of modern tractors, irrigation systems, and grain storage bins left thousands of farmers hopelessly in debt. And this, of all the ways that failure lurks on the American farm, may be the most lethal: raised on Protestant work ethic and a faith in the basic fairness of the system, most farmers firmly believe that the greatest success belongs to the family that works the hardest. The way out of debt is putting in longer days in the field, longer nights at the books.

But it isn't really so. In fact, at the end of the Farm Crisis, Danny Klinefelter, an ag extension economist at Texas A&M who studied and categorized farm failures during that era, found that the most common causes of the individual bankruptcies that eventually became an industry-wide collapse were: too much ambition, too much accrued debt, and, in Klinefelter's words, "too much wishful thinking." He said farmers didn't consider the cyclical nature of farming. They took on loans for land and equipment during periods of good weather and high prices. And if hard times fell, they expected to simply make up the difference with more work and more yield. It's rarely so simple.

"Many of the producers who have failed or are in trouble have been considered by the farming community to be top farmers," Klinefelter wrote, "but attaining the highest yields does not necessarily result in the highest profits."

That's a lesson that few farmers can accept. And before long, the cycle was repeating itself. In the decades since the Farm Crisis, grain production has kicked into overdrive. Big producers just got bigger. And small farmers, in an effort to keep pace with the expansion and vertically integrated models of agribusiness, began to take Butz's motto to heart. With the help of emerging technologies, everything from the GPS-mapped furrows to computer-controlled irrigation systems, they began to plant crops (especially corn) in places no one would have dared waste seed, much less water, a generation earlier. The more they planted, the more they stood to profit. But then, in the last decade, all of those acres and all that overhead started to become a curse.

As the current crisis unfolded, everyone with skin in the game—from the market watchers and commodities traders to the pundits and ag reporters (as well as every farmer at the counter of the café in every small town across the Corn Belt)—had his own explanation for the radical swings in the prices of commodity grains.

Some blamed reckless gambles by farmers themselves, and it's true that farmers take risks. They plant the shortest-season hybrids they can. They wait as late as possible to irrigate. And when prices were high, they planted every arable inch of soil, tearing out windbreaks along field edges and the brush along creek banks. For a while, the ditches on either side of every country road seemed to be stacked with slash. Farmers took land out of the federal Conservation Reserve Program, a sure payday for doing nothing but letting the ground sit, and they planted it to corn. The more farmers planted, the more they stood to profit, right up until everyone was drowning in surplus and the prices took a dive.

There are some old-timers who take the long view. They say that the consolidation of farms, going back to the Farm Crisis, puts decision-making responsibility into fewer hands, so that foolish risks are amplified. In the 1980s, most cropland was farmed by families with fewer than 600 acres. Today, the average farm is twice that size—and many are much, much larger. For more conservative farmers, wary of risk or simply too small and cash-poor to invest millions of dollars in acreage and equipment, just sitting still and doing nothing can leave you with less and less each year. Megafarms, with lots of resources, can afford to put more inputs on a field, to irrigate and fertilize and spray against pests. So in good years, when everything goes right, their harvest is high and drives down prices, which is fine for big guys with enough volume to profit from narrow margins but hell on the family farms. And in years when prices climb on some disaster, small farmers can still lose out by not having enough acres to capture the gain.

And, of course, there will always be those who blame the government. They say these big-time risks are encouraged by federal crop insurance and farm subsidy programs. It's true that farmers started taking out more crop insurance when Bill Clinton's administration made modifications to the Federal Crop Insurance Act and the USDA took over paying premiums. And it's true that more than $277 billion was allocated for farm subsidies from the expansion of the program in 1995 until 2014—with roughly 75 percent of the money going to the top 10 percent of farmers. So they raked in the majority of profits and set the market prices, while nearly two-thirds of American farmers struggled, collecting no subsidies at all.

All of these things are true, and they feed on each other, but none of this is really what caused the market to peak in 2005–2007—or to crash to its lowest point in decades. The problem was really a side effect of good intentions by the U.S. Environmental Protection Agency (EPA). In an effort to reduce carbon emissions and improve air quality, the EPA imposed the first Renewable Fuel Standard in 2005, requiring the production of at least 7.5 billion gallons of renewable transportation fuels within seven years. The goal was to kickstart the biofuel industry, but the incentive, when combined with soaring gas prices due to the Iraq War, instead triggered runaway demand for ethanol. The number of ethanol plants nearly doubled overnight. At first, that seemed like a good thing. It created a domestic fuel source, and the higher prices profited American farmers and the American companies that support them.

Soon, however, so much corn was diverted into ethanol production that it created a scare on the global commodities markets. Foreign countries dependent on American grain worried that their own farmers wouldn't be able to afford to feed their livestock, sparking a commodities run. The per-bushel price of corn shot up from $2 to more than $3. Everyone wanted to stockpile grain as a hedge against even higher prices, but American grain reserves were at the lowest on-hand level since the drought years in the Clinton era. After more than three decades of corn prices hovering between the $2 and $3 mark and generally trending downward on improved production since the early 1970s, prices shot up out of control. Depending on whether you were a corn grower or a livestock producer counting on that feed, the *New York Times* wrote, "you have daydreams—or nightmares—of that $5 mark."

In fact, from 2005 to 2008, the world price of corn broke past $5 and just kept climbing. And as corn reached unimaginable heights—$6 and $7—demand increased for soybeans as a feed alternative. Soon those prices had doubled too. As great as the run was for American farmers, food shortages and ensuing riots were touched off in parts of Asia, Africa, and Latin America. And the panic was enough to spark a run in commodities speculation. Traders had been looking for somewhere to put their money now that the securities markets, which had been backed by high-risk, low-interest home loans, were in free-fall due to the mortgage crisis. That money now pouring into the grain market created a brief price correction, but then the shortage brought on by the drought of 2011 sent commodities to record levels with corn reaching as much as $8 per bushel and soybeans topping $17 per bushel.

But on the farm, this is failure's best hiding place—smack in the middle of runaway success. One afternoon, sitting at the workbench in Rick's barn, he told me that the biggest pitfall you face as a farmer is your own optimism. Everything starts to seem easy, and you think the good times are going to last forever. And the government, in the name of spreading the wealth and stimulating the broader economy, always provides the excuse to take on more debt. In 2010, Congress announced that they would raise short-term rapid-depreciation write-offs for farmers from 50 percent to 100 percent, for one year only, so now, if you bought a new tractor or a new grain bin and spent $100,000, you could put that full amount against your annual farm income, instead of only half.

"People were just buying equipment like crazy because you could take that all right off of your income," Rick told me. "If you were in the thirty percent bracket making a hundred fifty thousand dollars, you could figure that you were going to give fifty thousand to the government anyway. But if you took a loan for a hundred-thousand-dollar purchase, the tax was offset." In 2011, across the country, farms suddenly had new trucks and tractors, new barns and outbuildings, but also new houses and new purchases of the kinds of equipment that can cost as much or more than a farmhouse.

"And here's the trap that farmers fall into and we're still dealing with. Like that big sprayer," Rick said, pointing to the vehicle in the corner of the barn, its wheels taller than me. "That was a quarter-million-dollar purchase." He paused a moment looking at the machine.

"As my banker says, people forget they have to pay the principal," Rick continued. "Now, during these really shitty times, everybody still has the payments. Yeah, I did benefit from all that tax write-off, and it was a hell of a deal on the interest too—but it's still an expense." Too much ambition, too much wishful thinking.

In no time, the factors that had quadrupled grain prices self-corrected. Gas prices stabilized and began to fall, lowering the market price for ethanol with it. Before long, ethanol producers began idling their new plants, because they simply couldn't produce fuel cheaply enough given the high input price of grain. As EPA tax credits expired, and the agency adjusted the renewable fuel standard downward, the drought also broke, bringing an upsurge of production in the 2013 harvest—just as demand was leveling off. The prices of corn and soybeans crashed, losing half their market value between the planting season and the end of that harvest year. So coming into the 2014 season, every farmer had to make his own market projection.

Rick gambled on another dry year and that the markets would remain down, so he planted mostly soybeans, which have been less volatile than corn, and he planted very short-season hybrids, hoping to get most of his harvest in early, before the markets fell even further. It should have worked—but in the high-stakes, narrow-margin game of commodities farming, a little rain is all it takes to derail everything. With his crops still sitting in the fields and markets continuing to fall, I asked Rick if he was worried—not stressed or anxious, but truly worried for his family. He shook his head. They were more than equipped, he insisted, to weather through a bad year.

"Now, if we see sub-four-dollar corn for two more years," Rick continued, "yeah, you'll see some people going broke, especially the young guys who started about five years ago when it was eight-dollar corn. And they thought, 'Oh, man. There's nothing to this farming game.' They bought the $12,000-an-acre ground and took out loans for the bins and the pivots and equipment.

"The Meghans and Kyles of the world," he said, "they could be in real trouble."

DARKNESS HAD fallen over eastern Nebraska.

After another day of chipping away, working and then waiting, Rick had decided it was time to call it an evening and sent everyone home. But he wasn't quite ready for the drive back or giving himself up to sleep. So he had ducked into the empty house on the southeastern corner of the Centennial Hill Farm, the house where his wife Heidi grew up.

It's abandoned now, except as a guesthouse and a place to host occasional potlucks. The refrigerator often holds a random collection of unopened cans and jars—not much, but enough to scavenge for a snack or a stray drink. Tonight, Rick scoured the remnants of the last get-together in search of something to take the edge off before bed.

"This bottle," he said, holding it up to the fluorescent bulbs overhead, turning it sideways to divine how much wine might be floating at the bottom. "I think it's been open for . . ." He trailed off, searching the label for a date before giving it a final shake. "Cleaning solution," he decided. With no other options available, he settled on the last two cans of Pabst Blue Ribbon but not without a few grumbles. He cracked one and took a long sip, then smirked. "I've raised hell about these in the past," he said, "but, you know, it's funny how you get certain things the right amount of cold, and they all taste pretty good." He slid into the booth of the kitchen breakfast nook and motioned for me to slide in across from him. He wanted to assure me that, despite everything that had been going wrong, he wasn't worried about the current crisis.

"I don't feel like it's going to be anything like the eighties when everybody was going broke," Rick said, taking another deep draw of his beer. "That was caused by Reagan and Volker, the Fed chairman. They thought they came in with a mandate to crush inflation—and inflation *was* out of control—but bankers had been pushing farmers to buy ground every year. It inflated up to about $2,500 an acre for good ground. The interest had been less than the rate of inflation. So when they raised rates, to all of a sudden have interest triple, that's what caused it. Bankers and farmers couldn't react fast enough. They went broke, had this huge debt. Bankers were foreclosing on farmers because your collateral was worth a third of what it was when you took out this loan."

He said that he'd witnessed this sudden decline firsthand. When he met Heidi, while she was a student at the University of Nebraska School of Technical Agriculture in Rick's hometown of Curtis in the southwest corner of the state, everything had been different. He had been a hotshot horseman in his small town—a skilled rider with a row of purple and blue ribbons in barrel racing and pole bending at the State Fair to prove it and still brash enough to take all comers in bareback sprint races on the ranches around Curtis or the dirt main street of nearby Wellfleet. He seemed to have the world by the tail. But college in Lincoln had been different, harder. After years of classes, he had taken a semester off and gone home, but he found life on the farm getting harder too, especially for a family like his that didn't own the ground they farmed.

Rick found himself at loose ends, as he took off another semester and then another. He was still technically enrolled at the University of Nebraska but a few credits shy of a degree in Latin American studies and in no hurry to finish up. The degree wasn't going to do him any good. He knew he wanted to be back farming and ranching, but he could see that the opportunities were drying up in Curtis. So he took odd jobs, while he figured out his next move. "I'd worked on the railroad for three years, steel gangs and tie gangs," Rick told me. "They'd made me assistant foreman and wanted me to be foreman. I could see then, if I didn't quit, I never would." So he left the railroad, but he still couldn't figure out what he wanted to do. "I just wasn't finding *it*," he said.

That's when he met Heidi. "Oh, everybody loved her," he said. "She could drive a tractor, fix engines." She had grown up in a farming family back in eastern Nebraska with a hard-driving father and three headstrong sisters. Rick's big-headed swagger didn't fluster her a bit. "We had a few dates. I even took her to church—very honorable," Rick deadpanned, then grinned, "but I was still convinced that I could change the world." He broke off the relationship and signed up for a two-year stint in the Peace Corps. He went off to Ecuador to teach agriculture in a small village, but he soon became disillusioned by the shortcomings of trying to apply American know-how to places with no resources or infrastructure. He came home disheartened and more lost than ever.

Rick paused, rattling his empty Pabst.

"Is there something wrong with your can?" I asked. "Some kind of beer deficiency?"

"I'm running low," Rick said with a smile. He cracked the second can. "And I'm just getting to the best part."

When Rick returned home from the Peace Corps, he found that everything that had been starting to slip away in Curtis was now utterly gone. Agribusiness was swallowing up small farms. Families had moved away. His hometown wasn't home anymore. "Dad had quit farming, and my sister and brother-in-law were farming that little rented place," Rick said. "I had grown up on a small rented farm. I knew there was no future there." A friend of his named Kevin had sold his farm and equipment and gone to Colorado to open a ski shop. Rick followed him out there and for a while worked a few hours a week and ski-bummed the rest of the time. At some point, he'd had enough. "I said I gotta get back to school and finish my degree."

And he'd been thinking about Heidi. He called her up to see if she wanted to have Thanksgiving with his family in

Curtis. After that, they could drive back to eastern Nebraska together—Rick to get himself reenrolled in school, Heidi to visit her own family. Heidi agreed and drove out to Colorado to pick up Rick, but by the time she arrived it was after dark and snow was swirling.

"That night we had a hell of a storm, about two foot of snow," Rick remembered. "She was driving a little Fiat." While he packed up the last of his stuff, his friend went outside to check on Heidi. She had put her farmer coveralls on and was under the Fiat putting chains on the tires. "Kevin came in and said, 'I think this one's a keeper. You better marry this gal.'" They drove across the state, all the way back to Lincoln through the storm, together. By the time they arrived, Rick had started to fall for Heidi. He worked on his degree, and on weekends, he drove out to York County and wooed her by pitching in on the farm. Eventually, he won her over, just as her father, Tom, was getting ready to retire. Soon, Rick dropped out for good, just a few credits shy of a degree, and started talking about leaving Lincoln, moving in with Heidi, getting married.

Audacious and brazen, Rick had a head full of ideas for ways to expand the operation and make way for the future. "In the 1980s," he told me, "inflation was so high that bankers were encouraging farmers, 'Hell, borrow money, because next year that money will value less.' So everyone took out loans against their farms. Then ground went from $2,500 an acre to $800 in two years, so farmer equity and the value of what their loans were borrowed against, went to nothing. And then the banks started selling them out." It was a terrible time for family farms, but Rick also saw the opportunity, the once-in-a-lifetime chance to leverage the land that Tom had carefully stewarded, in order to pick up more ground at a fraction of its real value.

Rick was still too young to appreciate it, but Tom himself had always been an early-adopter, a firebrand and risk-taker in his own right. But these times had him spooked. Tom's father had been put off the land when he was young, his grandfather forced to sell it off after bad decisions in the teens. His father had worked his whole life to save up enough to buy the farm back, and when he died young and unexpectedly, Tom had had to give up his own dreams. After just one semester at the university, he dropped out and came home to save the farm from being lost again. Whatever the ambitions of the young man who wanted to marry his daughter, Tom was reluctant to make unnecessary and untested changes and even more cautious about taking on debt in the midst of a credit crisis.

"But when Heidi and I got married, and she inherited her share of the land," Rick said, "I was fifty percent of the decision-making process on how to go forward." And he couldn't bring himself to pass up the opportunity he saw before him. Heidi had 50 cows, 50 sows, and 300 acres of land under her control. "When I first came back from the Peace Corps, I was going to make it on forty acres, and be all holistic and symbiotic, and everything working together," Rick said, but something about having more land and livestock than he'd ever imagined possible, along with a desire to prove himself to his father-in-law, brought out his competitive side. Rick and Heidi worked it all out on paper and decided to take the risk. They upped the operation to 150 sows and 100 cows. Heidi's father thought they were being reckless. "We had a hell of a good year, that first year. We worked our backsides off, but we made a hundred thousand dollars on hogs." To free themselves from Tom's second-guessing, Rick and Heidi used their profits to buy out all of the remaining debt on the equipment. It took every penny, but their piece of the farm now belonged to them outright.

But it didn't stop there. "After two years, I got Heidi talked into taking on three more quarters." Now approaching the peak of the Farm Crisis, land prices were at an all-time low. Rick was eager to buy up as much as possible, rolling each year's profits into buying still more land, on the assumption that an eventual recovery would set them up for life. So just three years after taking over their share of the farm, Rick and Heidi decided to use part of the original homestead as collateral for a loan to make still more land purchases. "I could see what a land base did," Rick told me. "Without land you cannot operate." He said he had everything nailed down for a loan from First National in York, but the day before he was supposed to sign the paperwork, he received a phone call from the loan officer.

"We've got a problem with our board of directors," the banker told him. "Because everyone's going broke, we want Tom to cosign."

"By then Tom and I were just fighting like hell," Rick said, "so I told him, 'Evaluate the loan on its own merits and if we don't qualify, we will seek assistance elsewhere.'"

Rick tipped his head back, draining the last of his beer. Then he lined up the two empty cans in front of him and shook his head ruefully. "So arrogant. Farming just three years, and I was that brash." The bank agreed to make the loan, and the pattern was set for the rest of Rick's farming life. "Because of my aggressive ways, we have continued to stay in debt for thirty-two years and always pushing, doing a lot more than we should. My idea was to try and build an operation that I could hand down to my children with as much land as what we were benefited when Heidi's share was handed down to her. So I did everything I could to get ahead and to turn that success into more and more contiguous land. If you can swing it, you buy it. When you're a farmer that land means everything."

It was late now, pitch black outside the window. Even the moon had waned to nothing but a slim sliver of light. Beyond the reach of city glow, the only sense of a world outside came from the distant barn light, its dim bulb always left burning and just bright enough to give shape to the shadows.

"I'm probably close to bipolar," Rick said at last. "When I'm on, everything is possible. I just go for it. But then I'll get really down, thinking, 'God, how are we going to get out of this?' And I put it on my kids, this guilt for my avaricious and aggressive ways—that I was doing it for them. That was my excuse in my head."

He sighed deeply. Months before, Meghan had warned me that one of Rick's defining characteristics is self-doubt. "He worries every decision to death," she said, "and then, no matter what he decides, he always thinks there could have been a better way. Rick is a very regretful man." It's a common trait among farmers. The neighbor's corn always looks taller, their cattle fatter. Every farmer kicks himself for not doing enough to capitalize when the markets are up and for being too exposed when the markets are down. Rick's mood, in particular, seemed to rise and fall with the prices from the Board of Trade. He regretted every missed opportunity but even more deeply regretted the times he hadn't been more cautious, more careful in planning ahead. Why hadn't he set aside more money when there was $8 corn?

"I just get comfortable, to where the wolf's away from the door," he said, "and what do I do? I go and remortgage everything and do dumb things—buy more land, more equipment, build a barn or a new house. It's like, *Rick, quit digging. You know what this is going to do to you.*"

He tapped the empty cans on the table in front of him.

"And it's caused a hell of a lot of stress on my family," Rick said. "Was that worth it? I don't know. That will be for them to decide."

KYLE DROVE the pickup south across the interstate and then back east toward what all the maps label as Lushton, Nebraska, though it's barely more than a wide spot in the road anymore. And we were still several miles outside of town, making our way toward Seth's farm. Kyle explained that Seth's wife is a Barlow, that his mother-in-law was Joan Barlow. "When we talk about 'Joan's half section,' that's who we mean," Kyle said. "Seth farms most of the Barlow family land, but we farm that one property." Kyle told me that he didn't know how exactly to explain why Rick and Seth had started harvesting together, but it had something to do with a kind of shared stubbornness.

A few years back, for example, Seth had taken a load of high-moisture corn to the grain elevator and been charged a drying fee. But Seth had seen the elevator simply blend his corn with a load of overly dry corn, so they hadn't actually had to turn on the dryer. "I was there," Seth told the operator.

"The dryer was not running. I am not going to pay a drying fee." Rick had told me earlier that he'd admired Seth for telling the co-op that it wasn't right. "I thought that was pretty gutsy of him."

As we hit the rise and the acreage appeared, we could see a wide field of ready soybeans and then a shelterbelt of tall trees, a farmhouse and cluster of outbuildings tucked behind. The beans, brown and mature, seemed almost to glow in the October light. "Wow," Kyle said, surveying the field. "It's ready to go." But then he turned along the south edge of the property and drove slow, so he could count rows and divide them to figure out the number of rounds to complete the field. As he went, I could see the shimmy and drift of Seth's rows. "He just hasn't invested in auto steer yet," Kyle said and laughed. "He just put this pivot up last year. Before that it was flood-irrigated with the pipes and all." It was a system that most farmers, especially in eastern Nebraska where center-pivot irrigation was invented in the 1950s, had abandoned long ago. Given the labor of changing gates, Seth was reluctant to irrigate except when it was absolutely necessary. It seemed to me that maybe Rick's admiration for Seth came from his ability to keep his farm producing by force of sheer will.

But it meant that Rick and Kyle were going to have to harvest the old-fashioned way—keeping an eye on the outside furrow, following the wobble and drift of rows planted by the human eye rather than computer-drawn straight lines. The mood, though, as we got out and Rick and Kyle readied the combine, was decidedly light. It was Meghan's birthday, and she was in the grain cart, chattering away to Rick and Kyle over the CB. Seth would swap in occasionally. He had his daughter with him, and she was having a great time taking the wheel of the John Deere. She was used to riding with goggles and earplugs in her dad's clattering old International Harvester, and she was sold on the air-tight, cooled cab, the sounds of Kyle's classical music playing over the radio.

But as the sun started to set, the mood seemed to shift with it. Because Seth had planted by eye, every turn of the combine required lining up with where he had tried to match row spacing between each pass at planting. Farmers call these "guess rows." The sixteen rows laid out by the planter are perfectly spaced, but the furrow between passes can be slightly narrower or slightly wider than the machine-spaced rows. Because the harvester is twelve rows wide, it's impossible to split each planting swath. Instead, you have some passes within the planted pattern and some where the harvester is straddling the places where two planting rows come together. But as the light failed and the shadows grew longer, it was getting harder and harder for Kyle to keep track of the guess rows and remain within the pattern. He leaned forward, peering down over the steering wheel, to be sure that the spinning reel of the harvester was standing up the rows and cutting everything cleanly.

Seth, back in the spring, had planted around the concrete platform of his new center pivot for the first time. And rather than setting his row-spacing starting from that platform, he'd started from the edge of the field and worked in. Ordinarily, farmers won't plant closer than 30 inches from their pivot, but Seth, hoping to get a little extra from the field, had allowed just 5 inches. So as they neared the platform, Rick and Kyle had to figure out where to set the floating bar on the outside of the harvester head, what they call "the snout," in order to stay aligned but without having to take multiple passes to get around the pivot.

"We try to put the outside of that head on the guess row," Kyle explained. "But the eye-planted rows can sometimes run together. And if you don't guess right, you're taking half a swath trying to fix it—which takes fuel and time." And if you get off by too much—you miss a guess row and don't notice or just try to keep going—you can start to run over rows, breaking the pods or treading the plants into the muddy soil and making them impossible to harvest. Seeing the trouble it was causing, Seth told them not to worry about the rows right around the pivot, but, even in a neighbor's field, Kyle wanted to capture every bit of the yield. So between passes, he would come down from the cab, study the rows with Rick and then hustle back behind the wheel.

At dinnertime, Meghan drove to pick up food Heidi had waiting for them at the house. Meghan returned with tacos wrapped up in aluminum foil and a cooler full of Cokes. Everyone sat on the tailgate of the pickup or leaned against the bed, eating quickly. Rick kept eyeing the sun, now sinking into the line of trees that stood between the edge of the field and Seth's house.

"I think we should call it a night," Rick said finally.

"It's all right," Kyle said. "There's just a few more swathes."

Kyle was hoping to finish this field before it was full dark, so they could load up the head and the combine for another field for tomorrow. If they could get done in the next hour, they would have a jump in the morning. Together, Rick and Kyle walked over to the remaining rows, getting a read on how many rounds were left to complete the field, but also going back and forth about whether to continue or stop. Finally, Kyle won out. He took the combine down to the north edge of the field, then turned back toward the pivot, watching the spinning reel below him and steadying the wheel.

As he neared the south edge of the field, Meghan, waiting in the grain cart, came on the radio. "You're driving crooked," she said. "You're knocking over rows." Kyle looked down at the reel. Everything seemed in line. So he stopped the harvester and climbed down to see what was going on. Right away he could see: the snout was bent in and the plastic body of the soybean head was broken. He'd hit the pad of the pivot and never even felt it.

"Goddamn it," Kyle rasped under his breath. This could be thousands of dollars of damage. Worse still, if it was more than he could fix himself, it could be days of waiting for a Deere certified mechanic. He pulled back the plastic body and he stuck his head inside. Kyle told me later that he could see that the snout had been pushed in and bent up the hydraulic arm on which it floats. That arm was rubbing on a pulley belt that runs the cutting sickles, putting slack in the system and creating friction. "It was still cutting," Kyle said, "but it was already getting really, really hot." He reached in to see if he could straighten the bent arm by hand. No dice.

"I should have just left that row," Kyle said, still tugging hard on that arm. "It was planted *right* up against the pad."

"That's what Seth said," Rick snapped. "You *cannot* harvest something that's planted that close."

"Yeah, I probably should have just left that little bit."

"If Meghan wouldn't have seen it, it could have started a fire," Rick said. He couldn't hide his anger, but Kyle was already thinking about what they needed to do now. He said he could straighten the arm if he could just heat it up, but he couldn't afford to wait until morning to fix it. If they were going to finish this field in the morning and still stand a chance of loading up and getting some beans out of their own fields, Kyle was going to have to make the repairs right then, in the dark. He asked Seth if he had a cutting torch and a wrench he could borrow. Seth told them to pull around to his shop, and he set off across the field.

By the time Kyle and Rick had driven up in the pickup, Seth had already rolled out a portable light and the tank and hose and a rosebud heating tip for the torch. He grabbed a couple of wrenches. Kyle asked if he had any fender washers. Seth pulled out a box and shook several into his hand.

Hoping to get out of the way, I went with Meghan back to the house, where we waited, saying little for close to an hour before Rick and Kyle came rolling into the driveway. They stomped into the mudroom laughing and kicking off their boots. "You asked for that fender washer, and Seth just said, 'How many you need?'" Rick said, slapping Kyle on the back. "Why can't I have a shop like *that?*"

"I assume everything's working," Meghan called from across the kitchen.

"Yeah," Kyle said with a long sigh. "We had to drill out some rivets and get the plastic bent up so we could get the torch in without burning everything up. And then we had to heat up that arm to get it straightened out so it quit rubbing. It was really hard steel and we couldn't get it bent by hand. And then a little piece on the backside broke. So we had to weld that back together."

Rick interrupted. "I want you guys to know: the combine is better than it's ever been."

"Better than before?" Meghan asked skeptically.

"At least," Rick said. "We should hit every center-pivot pad."

Before long, they were seated together around the dinner table, recounting stories of all the near-disasters of past years. The time Dave took out a power line with the unloading auger. The time Meghan had swung wide to make a turn on a country road with the bean head strapped onto the flatbed trailer and clipped a stop sign. They laughed until all the tension and worry of the accident, the adrenaline of what could have happened, had been shaken off and disappeared.

In the end, American farmers harvested more than 14 billion bushels of corn in 2014 and nearly 4 billion bushels of soybeans—setting new records, as they had the year before and would hope to in the year to come. With so much production, roughly three-quarters of the harvest nationwide went directly into bins, as every farmer waited and prayed for rebounding prices. They never came. Instead, prices continued to slump as yields continued to grow, and whispers spread about the possibility of another Farm Crisis. But for now, on that one night in October, the last hours of Meghan's birthday nearly gone, everyone was gathered in the kitchen, happy and laughing until they were ready to sleep and go again in the morning.

"The hardest part of my job is working with family," Meghan told me later, "but that's also the best part of my job, because family can be pretty hard on each other, but at the end of the day, they're the ones that will be there for you in the hardest times. We've been through some hard times on the farm, and we're still here, still going—and hopefully on to the next generation."

WORKING COWS

November 2014

First the truck appeared, then the cattle trailing along behind. The November sky was slate-gray and cloud-strewn, the temperature holding in the forties, just warm enough that the intermittent rain came as a dry mist rather than snow. Meghan was at the wheel of the white pickup, slowing occasionally to bang on the side door and call out to the hesitant cows. Their big bags swayed under them as they ducked over the furrows. "Come, boss," Meghan called, beating the door again like a drum. Behind her, Rick and Kyle on four-wheelers stitched along the edges of the herd, circling up stray calves and keeping the line moving. Kyle, in a stocking cap covering his buzz-cut head, stood up on the running board, straddling his seat. He kept his hands on the handlebars, still easing forward, as he scouted across the line of jogging cattle. If he spotted a loose calf or errant cow on the other side, he whistled to Rick, who zoomed ahead, shouting in a hoarse staccato. "Go on, now. Go, go, go. Go on, get moving."

After the end of harvest, when the last of the beans were in the bins and the calls for white corn had started to come in from the elevator, when the rented harvester had been loaded back onto the flatbed and sent on to some other farm farther north, Rick turned his attention straight to getting the cattle ready for winter and then for spring calving. He drove the trailer out to Curtis, where he grazed the spring cow-calf pairs on pasture, growing them on grass through the summer and start of fall, and loaded them up. He brought them back to York County and ran them onto a quarter section of freshly harvested corn surrounding the solar-powered barn.

In fields like those, where the corn has been well irrigated by a center pivot, the cut stalks standing in a winter field along with the wayward cobs and kernels scattered by the combine, what farmers call "crop residue," can provide ample feed without any additional cost. The moisture left in the stalks plus the greens from the turnips and radishes planted as winter cover crops are enough to get lactating mothers through until weaning—and for those cows without calves, calories from those cover crops can actually help them put on pounds. But if the cows were going to be nursing newborns in a few months, then it was time to separate the spring calves from their mothers in order to wean them, and it was time to check the cows for new pregnancies.

Meghan groaned just thinking about it all. "When harvest is over, you'd think you could catch your breath," she said. "But we're straight into the cows—bringing them home from pasture, sorting them, weaning them, and getting them to the sale barn. It's a lot." She hit the gas, speeding up to the pipe corral attached to the old barn. She jumped out of the truck and swung open the wide metal gate. "C'mon, girls," she called again. Heads bobbing, the first of the mothers walked obediently into the pen, their calves skipping but sticking close. The Hammonds raised all Angus cattle for beef. In recent years, Rick had moved away from hormones and feed-grade antibiotics, and he had tried to do more direct marketing, selling halves and quarters to customers willing to pay a premium for hormone-free, grass-finished beef. But with so much risk already sitting in his grain bins, Rick was content to get his calves to market and get more on the way. Beef prices were soaring, hitting $6.30 per pound at retail, more than a dollar above where they'd been just a year before. Most years, Rick regards the cattle as a side operation, work he enjoys but doesn't expect to pad his wallet. But with prices climbing and several years of drought driving the national cattle herd to a lower point than it had been in decades, the sale of beef promised to be the farm's saving grace for 2015.

Once the cattle were all in the large holding pen, Meghan swung the gate closed and grabbed her notebook. They would go systematically through each cow-calf pair, checking ear tags and noting any obvious health problems, before running them into separate pens. After that, they would herd the calves into the chute for more detailed health inspections. Rick, Meghan, and Kyle, as well as Dave, who had been waiting back at the barn, all filed into the wide paddock, the cattle stooping and scattering around them like startled birds. Rick had a pair of rattle paddles, like two oversized lollipops, lemon and orange, but inside those bright-colored plastic pads were ball bearings that shook and spooked the cattle forward.

Arms stretched wide, coaxing and calling, shaking one

paddle and then the other, Rick separated calves, one by one, from their mothers. He sent the weaned calves one way and then the other, until they skittered toward the gate, which Dave would swing open to usher them into the calf pen. If they showed signs of balking at the doorway, Dave and Kyle had herding rods, nothing more than long flexible poles, to tap their hindquarters—just enough to trigger a prey animal's flight instinct and send the calf galloping ahead. Eventually, when the calves were separated out, the mothers were moved as a group to another pen, and the next batch of pairs was brought in.

This was how it proceeded for hours, cajoling and cursing. The four dust-covered and wind-chapped humans wading into galloping herds of erratic cattle, their black flanks twitching and mud-patched, their hooves kicking up more clouds of dirt in the holding pen. Rick would call out to the cows, then to his helpers, and before long he was yelling at Meghan, Meghan yelling back at him, and then the two of them were laughing at each other. And with each new group, the din of the corral grew: calves bawling for their mothers, worried cows mooing in response, straining to lift their mouths, tongues lolling, over the crush of other fussing mothers.

"Working cattle is a high-stress time," Rick had warned me ahead of time. "Be prepared for a lot of noise." There was certainly that, but it only dawned on me later that it wasn't so much the chaos he was worried about as how I might respond to the distressed and calling mothers or the sight of the skittish calves as they bolted through the gate and were snapped into the chute, so Meghan could complete a wellness check, looking for tumors or signs of nasal or bronchial infections or pinkeye. She marked down the information for each as she went, the pinned calf tense in the chute, sometimes letting out a meek bawl, sometimes releasing a milky panicked poop, its eyes rolling and wild to be free.

"We had pinkeye terrible this year," Meghan told me later, "and pneumonia in one group." She shook her head in frustration—another side effect of all the rain late in the season. "We don't get pinkeye out west, because it's a drier climate." But in the eastern part of the state, flies hatch out with the rains and swirl over manure piles left in the fields and feed yards. When those same flies cluster around the eyes of cows, tonguing the salt right out of their tears, pinkeye can spread and then pass from mother to calf and through the herd. "We treated them all ourselves, except for a few that were pretty bad," Meghan said. "They had to go in and get a shot in the eye and have the lid stitched shut." That subconjunctival injection of antibiotics to kill the infection, mixed with steroids to knock down the inflammation, usually takes care of the problem in just a few days, but it also means that the animal can no longer be sold as antibiotic-free or steroid-free. A rash of pinkeye can close the door on a whole season's direct marketing and its premiums.

This second-guessing of basic animal husbandry, often to satisfy a grocery shopper who has never been on a farm, is a sore point for almost anyone engaged in raising livestock these days. Every cattle rancher is quick to talk about how they have scaled back on chemicals and drugs, and they can barely conceal their suspicions of outsiders, who might object to ear tags or branding, or mistake commonplace herding for rough handling. Even in the intricate systems of corrals at livestock auctions, where most cattle are still moved by skilled cowboys on horseback and drovers with herding rods, I've been told innumerable times that electric prods are rarely necessary, that no one wants to stress the animals. At one large feedlot in Bridgeport, Nebraska, the owner wanted me to know that his son had gotten a bachelor's degree from the sustainable agriculture program at Yale and that his daughter had gone to college at Colorado State, where she studied under Temple Grandin. She had come home and redesigned all of their chutes and gates, he told me. Almost across the board, beef producers feel scrutinized and critiqued, and they have a love-hate relationship with a consumer base willing to pay extra for an antibiotic-free steak without any thought for the potential suffering caused by denying drugs to the cow that produced it. More than once a rancher or feedlot owner has asked me: "Would you prefer that I withhold medicine from a sick animal?"

By evening, when the gray sky deepened to black and the temperatures dropped below freezing, Meghan and Rick seemed fatigued both by the physical exhaustion of the all-day work and by the inescapable feeling that they were somehow under surveillance. As direct marketers of their own beef, they know well that much of the new food movement is built on a nostalgia for a bygone era of family farms and natural food production, and all of the farming magazines and websites are always urging them to "tell their own stories." They're not trying to hide anything, but they often feel forced to prove their heightened commitment to land stewardship and animal welfare and building sustainable agriculture in support of communities and health. And they know all too well that this urge for "food with a story" has only encouraged corporations, with much bigger marketing budgets and an army of branding experts, to reposition their products and remake their image. Not only have the big meat producers taken over any organic or hormone-free company large enough to take a bite out of their market share, but they have often assumed the identities of the very farmers they are trying to elbow aside.

In just the last decade or so, Smithfield Foods acquired Farmland, Tyson took over Hillshire Farm, and Hormel bought out Farmer John. Now, the products from some of the largest and most industrialized food producers in the world are packaged with logos featuring red Dutch gambrel barns, white farmhouses, and smiling farm families, decked out in denim shirts and overalls, holding grandchildren and hand tools for hoeing rows. Shoppers see those pictures and assume that's how their food was raised. "But how does the little guy get the message out?" Rick asked. He was slumped over the bed of the pickup, now wearing a brown pullover stitched with the logo for R Diamond, the name of his cattle operation. Consumers make demands, he said, and it's easy for major corporations to pass along those new requirements to their growers. "No sweat off their back," Rick said. But for farmers maintaining narrow margins and getting squeezed from both ends, it was hard not to feel trapped.

Later, as we all sat around Rick's kitchen devouring a pizza that Heidi had picked up at the Casey's General in Aurora, Rick downplayed his worry. "I don't care if I come off as the dumbest SOB," he said. "If there's truth in advertising, I don't see how it could be any other way." He said he was more concerned that life on an American farm isn't all it's cracked up to be, and that he appreciated it when people like Willie Nelson and Neil Young ride in and give concerts, but he had grown leery of how they romanticize small farms. "I'm afraid what you're going to find," he said, "is that the idyllic family farm that Willie and Neil have been trying to save does not exist. And by the way: the good old days weren't so good either. Financial woes and danger and tired all the time."

"I want you to tell the truth," Meghan interjected. "But you have to understand: our family and neighbors are also our coworkers. They're going to be there when someone dies or when we have a big fire or when the cows get out." She had seen how protesting against the Keystone XL pipeline and the spotlight of the media had turned one neighbor against them already, and she was worried about what further attention might do. "If there was something that our neighbors didn't like and they wouldn't let us farm their ground or run our cattle on their fields, I would just be heartbroken."

Meghan hesitated, then seemed to wave it all off. She put her plate and fork in the sink, the darkness outside the window now deep and complete. But then she had one more thing to add. She wheeled and braced herself on the edge of the sink, as if she'd suddenly been overcome by great exhaustion. "It's just that we've sacrificed so much," she said. "We work hard just trying to keep our family together."

PART TWO

THE HOMEPLACE

November 2014–February 2015

To understand, first remember: Nebraska is a place. It sits square as an anvil in the center of our maps, and yet, somehow, everyone manages to forget it exists. Maybe that's because Nebraska is also a land of ghosts, of small towns dwindling to the point where, in another generation, they might simply cease to exist. In Benedict, the nearest town to Centennial Hill Farm, the school stands empty and abandoned; the only restaurant has been for sale for years. There's a grain elevator, two well drillers, a feedlot outside of town, but, otherwise, there's no work, nothing to do, no reason to be there instead of anywhere else. West of town, approaching the hard edge between York and Hamilton counties, the roads lose their Civil War names—Lincoln Avenue and Shiloh Street, Sherman and Atlanta, becoming just a grid of letters and numbers, unfolding in perfect one-mile squares. At the crossroads of 22 and G, five miles west of Benedict, is Centennial Hill, the piece of ground held by—and holding—Meghan's family for generations.

It doesn't look any different than the miles of flatlands surrounding it in all directions, but it is the center of the family's universe, the homeplace. Every farm family has one. Though nearly all operations have expanded in the decades since the Farm Crisis, acquiring land and growing piecemeal, these are just scattered acreages, constellated satellites orbiting the homeplace's center of gravity. In most cases, the new land has no emotional importance and doesn't even get a new name.

So when families like the Hammonds talk about the patches they own and farm, it is a kind of geography of the gone: the Metz place, Roger's dry-land patch, Karen's quarter. It's all the people and families that disappeared off the land, leaving only their names, like tombstones, as a record of the generations spent there.

These places are critical now to the survival of the Hammonds' operation, but the homeplace, Centennial Hill, is different. To really understand why, to count the generations of hardship and hard work that have gone into holding on to the farm, Meghan said I should go see her aunt Jenni. She is the acknowledged keeper of Harrington ancestral lore. The basement of her Victorian farmhouse, cattycorner from the homeplace, holds the family archive—the scrapbooks filled with faded clippings fixed in place with yellowed and brittle tape, the box of old photographs, their corners all curled and chipping. When I arrived, it was cold outside, no snow on the ground but the winter sun pale and sharp. Light poured through the north-facing window into the dining room, where Jenni had laid everything out on the table.

It was uncanny how much Jenni looked like Heidi, like all of her sisters. They were all in their fifties but had an intangible youthfulness about them. They all had kind and lively eyes, a soft but frank tone of voice, and smiles that could be slow but genuine when they emerged. Jenni welcomed me inside and got me a cup of coffee. Once I had warmed up a bit, she brought me back to the dining room, where she sorted through stacks of sepia and silver prints until she found the one she wanted: a picture of the first proper house on the farm. Her father had had the place torn down when Jenni was still little, but it seemed to float in her memory, blurred and overexposed like the ghostly image in the photograph. It was irretrievable but also forever present. She studied the image, holding it up to the light in her strong, tan hand.

"Our great-great-grandfather was Thomas Barber," she said. "He came from Suffolk, England, with his mother and father when he was about age four." The family settled in Wisconsin in 1852 and farmed there, but it was less than a decade before their adopted country was divided by war. When he was still not quite sixteen years old, Thomas signed up for the Federal Army and was assigned to the 13th Independent Battery, Wisconsin Light Artillery. He had already crossed the ocean and traversed half his new country by rail; now he was sent the length of the Mississippi River to Louisiana, where he spent eighteen months as a member of the fortified defenses in Baton Rouge and New Orleans. After the Confederate surrender, Thomas returned briefly to Wisconsin but felt restless and eager to strike out for the territories. He was not yet eighteen when he went west.

According to Jenni, Thomas was hoping to take advantage of the Homestead Act. The first Homestead Act had been signed into law by Abraham Lincoln in May 1862, just as the trenchant and gory battles of the Shenandoah Valley Campaign in Virginia were signaling that the Civil War would be more than a brief insurrection. As the bloodshed increased, President Lincoln began looking not only for tactics to hasten the war toward conclusion but also ways to create a lasting postwar peace, both free from slavery and from the conflict the "peculiar institution" had created since the nation's founding. Lincoln's ingenious solution, the Homestead Act, granted small farming plots in the western territories to any citizen who had never taken up arms against the Federal government and was willing to occupy the land and "prove it up" with a year-round habitable home and a crop that could be shown to be sustainable for five years.

Bringing families onto the barren prairie might seem far removed from the battles raging in the Eastern Theater, but the dispute over western settlement—whether the Kansas and Nebraska territories along with other points west would be divided and designated as new slave states or free soil— had been the central political question of the 1840s and 1850s. The admission of Kansas as a free state in January 1861, just before Lincoln's inauguration, was the match that finally lit the fuse of civil war. It had realized the abolitionist dream of the martyred John Brown and, coupled with the fear that the new president was a silent opponent of slavery, it had propelled Southern states toward Secession.

But Lincoln, already a shrewd and hard-nosed statesman, recognized that the roiling of the country toward constitutional crisis had also created a political opportunity. With the South removed from congressional debates, Free Soil members of the House and Senate had the numbers to configure western settlement as they had always envisioned, as a stronghold of small-plot farms worked by free white labor instead of large plantations held by small numbers of wealthy landowners who thrived on the backs of slaves. Though he had only recently emerged as the unlikely leader of the Republican Party, Lincoln readily understood that such a realignment of the nation's social structure was both the mandate of his election and the best chance his newly formed party had for broadening and shoring up its base of power. The rapid settlement of the American West into small, self-sustaining plots would ensure that, when the horrific battles of the war had ended, the country would be firmly established as a free nation of independent farmers.

So Lincoln's team came up with a simple plan: give away the unsettled territories 160 acres at a time. The size of the allotment may have seemed arbitrary even at that moment, but it was actually the direct legacy of the Public Land Survey System. Under this plan, the western regions ceded to the new republic by King George at the end of the American Revolution and later the open prairieland of the Louisiana Territory acquired from Napoleon had been mapped out according to a rigid grid and tamed into townships, each six miles square, each township further subdivided into one-mile sections, and each section divided again into 160-acre quarters. Railroads stops were extended to each township, dirt wagon paths were built along the section lines, and farms were allocated one to a quarter.

With the passage of the Homestead Act, the grid was extended onto the flat expanse of central Nebraska, dividing the land evenly, eschewing the realities of terrain in favor of the harsh simplicity of geometry. In those days, there was no one to lend a name to the country lanes. And to this day, the east-west gravel tracks have only numbers; the north-south roads are lettered. The intersections are still coldly determined—precisely a mile apart, unchanged from the nineteenth century. The only deviations on the unbroken straightaways come in occasional hard jogs inserted by surveyors to correct for the curvature of the Earth. Otherwise the grid is only interrupted in the precious few places where it is slashed by the braided diagonal of the Platte River.

Despite this strict adherence to the vagaries of the plat maps, Lincoln was a pragmatist when it came to meeting the challenges of farming on the windswept prairie. Having spent years as a young lawyer and congressman on the frontier of Illinois, he knew life for homesteaders would not be easy; he understood what it would take to lure them and also what they would require to survive. So along with passage of the Homestead Act, Lincoln created the U.S. Department of Agriculture and also signed the Pacific Railway Act and the Morrill Act, named for Representative Justin Smith Morrill, who introduced the legislation. The Pacific Railway Act authorized large land grants to the Union Pacific Railroad and Central Pacific Railroad to encourage their growth westward. The Morrill Act provided each state with still more land grants of 30,000 acres per congressional seat. Funds from the sale of the land were to be used to establish schools of agriculture and mechanical arts, dubbed A&Ms. And the newly designated Department of Agriculture announced that it would conduct research on best farming practice and dispense information through the A&Ms to aid the growth of agriculture in the West through hard science.

The creation of the USDA was especially significant, because it heralded a new era of government intervention into agriculture—not merely providing incentives for spreading farming into the West but actively providing support. To head the new department, Lincoln selected a man with the unlikely name of Isaac Newton, a Pennsylvania farmer and entrepreneur who had risen to become superintendent of the Agricultural Division of the U.S. Patent Office. Newton declared that the USDA would focus on several key areas, including scientific analysis of farm data, collecting and analyzing soil, supporting the creation of hybrids and regionally specific crops, and testing and promoting new equipment. He set up an experimental farm on the National Mall to keep his initiatives constantly in view of Congress; he lobbied for funding to establish professorships of botany and entomology at the A&Ms; and he advocated for implementing British meteorologist Robert FitzRoy's use of the telegraph to gather daily weather observations, what he called "forecasts," that could be shared with farmers by way of their local newspapers.

By the time Thomas Barber arrived in Nebraska in 1866, the country was entering a renaissance of modern farming. But before he could take advantage of the nearly free land and agricultural know-how on offer from the government, he first had to figure out how he would meet the conditions of the grant. If he was going to occupy the land and also improve it with a home and a successful crop, he first was going to need some money. So in June 1866, he took a job on a farm in Sarpy County on the bustling edge of Omaha. He worked there for four years, seeing Nebraska become the first state admitted to the newly preserved Union and slowly putting away $190. He used those savings to rent a farm and built his nest egg for another two years. Somewhere along the way, Jenni wasn't sure just when, Thomas married Mary Mitchell and had a daughter. With a family now, and enough saved to make a go of things, he set out looking for land in the counties west of Omaha, particularly York and Hamilton counties, where the richest soil was reputed to be. But by now it was 1874, and he found that whole area had been homesteaded and much of the desirable farm ground there, which had been available for just $18 per quarter section barely a decade ago, was now selling for a steep $20 per acre.

That summer, Thomas caught what would prove to be an unlikely break. After nearly two months of drought in May and June, a disaster of apocalyptic proportions befell eastern Nebraska—and most of the American West. "In a clear, hot July day," one eyewitness in Seward County remembered, "a haze came over the sun. The haze deepened into a gray cloud. Suddenly the cloud resolved itself into billions of gray grass-hoppers sweeping down upon the earth. The vibration of their wings filled the ear with a roaring sound like a rushing storm." For ten days straight, innumerable grasshoppers swept across the middle of the country from Texas to the Dakota Territory, devouring everything in their wake. Newspapers reported that the locusts chewed through fields of corn so fast that it sounded like a crackling fire, and the clouds soon grew so thick that they stalled a Union Pacific engine at Stevenson station, near Kearney, Nebraska. By August, there was noth-ing left, and thousands of farmers, who had been counting on their harvest to keep them afloat, were panicked.

One of those early settlers, Reverend C. S. Harrison, had come to York County in 1871, in search of unsaved souls on the godforsaken plains. He had purchased a piece of ground and laid out a hatchwork of homesites. The railroad was rumored to be on its way, and Harrison planned to sell homes to the lineworkers, feed-mill employees, and shopkeepers who would follow. In order to receive an additional quarter section awarded by Nebraska under the Timber Culture Act of 1873, Harrison had planted trees along either side of the lanes— Cottonwood Street, Box Elder Street—hoping they would one day spread and arbor the yards. He predicted that the Great American Desert would soon be forested to the base of the Rocky Mountains and, to show his confidence, named his tiny tree-lined town Arborville. But that savage summer, the grasshoppers ate the paint off his church, the handles from plows, and crops out of the fields. They chewed the leaves, the *limbs*, of Arborville's trees down to the trunks.

Nearly three-quarters of the American agricultural crop of 1874 was lost. One Nebraska newspaper mused: "We have heard some sighing, 'Oh, for a Moses to lead us out of this land of grasshoppers.' We fear the next cry will be, Oh, for a Joseph to sell us corn." And, indeed, farmers soon had no feed for their cattle or hogs, and the bodies of dead grasshoppers choked and putrefied their ponds and open wells, leaving them without water. By October, with winter approaching, desperation had begun to set in. The *Omaha Daily Republican* estimated that more than 10,000 people (in a state of fewer than 200,000) were in need of help "to keep them from actual starvation, until next harvest." Those who feared they couldn't make it that long flowed back east, selling land for pennies on the dollar or abandoning it altogether. Hamilton County, hoping to stem the hemorrhaging, offered low-interest loans of up to $200 for anyone willing to stay, but still, families loaded their wagons and left.

Farmers will tell you that rural communities unite at moments like this, rallying together to survive. It's not really true. In times of crisis, the pragmatic and the prudent pack up and go. Only the most stubborn remain, over the years and decades giving rise to loose, often unincorporated and thus ungoverned communities of hardheaded hermits, prideful in having outlasted their neighbors. And they are experts in finding opportunity in others' misfortunes. Thomas Barber, with almost as much saved cash as any man in Nebraska could get on loan, arrived in York County and selected 120 acres of prime land, two miles east of Arborville—what he later named the Centennial Hill Farm. The hill is really something less than a rise, just enough to give the land a southward slope. So the property is sufficiently level to hold a steady rain, but not so pancake flat that the rows puddle in a downpour. Instead, runoff pools into a natural catchment along the edge of the acreage, bounded by the roadbed. It's a perfect spot for farming—easily planted and with a ready source of water.

Jenni handed me the family scrapbook, open to a clipping from some long-forgotten Nebraska newspaper. The editors had asked Thomas late in his life to recall his arrival in York County. "On the 13th day of November, 1874, I moved on to the land," he wrote. He had an 80-acre homestead and 40 acres picked up for cheap. He brought with him, in addition to Mary and their infant daughter, two cows and twelve hogs, along with the most modern farming equipment of the day: a combined reaper and mower, a threshing machine, two wagons, and six horses to supply the power. He built a sod house from the prairie soil, and almost exactly one year later, Lela, his second child and the first born on family land, was delivered there.

Over the next few years and successive summers of locusts, farmers continued loading up and fleeing Nebraska. As they did, Thomas registered neighboring fields as abandoned and bought them, and he filed a timber claim that allotted him more land in return for planting trees. His family's first farmhouse, the replacement for the original soddy, was erected in what would become a windbreak created by that line of even-spaced saplings, and its frame and shingles, its clapboard siding and floors, were built from the milled planks and beams of Arborville's denuded trees.

For the next six years, Thomas raised wheat and barley,

hauling it some thirty miles to market in York, and he picked up extra money by threshing wheat for his neighbors as well. "Just as fast as I could get any money ahead I put it into land until I secured 320 acres," he remembered. When the town of Bradshaw was platted along the Burlington & Missouri River Railroad in 1879, Thomas had a ready way to get his livestock to Omaha, Kansas City, and even Chicago. He expanded his herds of cattle and hogs and began growing more and more corn to feed them. Eventually, he was raising as much as two grain carloads per year, enough to start a small cattle-feeding operation. At the same time, he bought up surrounding plots of land, paying $20 to $28 per acre, until he had 520 acres of contiguous land.

When the Kansas City & Omaha Railroad arrived in 1886, the town of Benedict was platted exactly five miles east of Thomas's land. In no time, the depot was a hive of activity with supplies for Sparling's General Store arriving daily and grain and livestock being loaded to be sold in Lincoln and Omaha. A schoolhouse was built, where Thomas sent his children during the week. And on weekends, the opera house hosted phonograph concerts, dances, and plays. Perhaps it was there that Thomas's daughter Lela met Henry Moore Harrington, known affectionately as Harry, whose uncle had put up the land for the town and still owned the farm directly across the tracks. It almost certainly was there that Lela and Harry's relationship, little by little, grew into a courtship. On New Year's Eve 1894, they were married, and Thomas Barber, after twenty years on Centennial Hill Farm, turned the operation over to Harry and moved into York to retire.

MEGHAN WAS never as caught up in the tradition of Centennial Hill Farm as her mother and aunts. Yes, she grew up there, within feet of where Thomas Barber's original farmhouse was built, but as high school graduation approached, she couldn't wait to get away. She had a storybook senior year at High Plains Community High, the consolidated school drawing kids from farms for miles around. Even in a tiny class, where everyone does everything, Meghan stood out. She was the homecoming queen, senior class president, junior marshal, band majorette. She played on the volleyball and basketball teams. But she had grown up skiing in Colorado, taking vacations to visit family in New York. For her, brighter things beckoned. When she went away to Metropolitan Community College in Omaha, she told me, she wasn't looking back. "I was sure that I was never coming back to this place," she said, "never going to scoop any more poop in the barn, never going to sort cows again in my life. I was done."

Her high school sweetheart, Brent Zoucha, had graduated a year ahead of her. From the small town of Clarks, just across the Platte River from the house Rick built when Meghan was in high school, Brent was the all-American kid. He worked at Pollard Oil in Clarks, changing oil and fixing cars, and he was a star of the basketball and track teams, placing in the top three in the high jump at the state track meet in both his junior and senior year. But Brent was also shy and a little goofy. Initially unable to work up the courage to ask Meghan out, he had stopped her by her locker to see if he could borrow her calculator. It was enough to get them talking, and eventually they were inseparable.

When Brent graduated in May 2005, he enlisted in the Marines, following in his older brother Dyrek's footsteps. Brent went through basic training at the Marine Corps Air Ground Combat Center in Twentynine Palms, California. Meghan had wanted to attend his graduation that January, but her parents insisted that she stay in Nebraska and focus on finishing her own school year. She was there in Omaha when Brent arrived back home after boot camp, and she was there again seeing him off from the airport when he was deployed to the Anbar Province of Iraq as part of the Marine Expeditionary Forces. They kissed and said their goodbyes. Being apart from Brent for nine months was more than she could imagine.

With Brent now gone, Meghan made plans for a late-summer trip to Ireland before classes started in the fall, and she spent the rest of the summer hanging out with friends in Omaha and Lincoln. On an early morning in June 2006, Meghan was sleeping at a friend's apartment across the street from Southeast Community College in Lincoln. "We'd been out partying the night before," she told me. The landlord called first thing in the morning, rattling their hung-over nerves, to say that there were going to be repairmen coming by later in the day. They had just gotten back to sleep when the phone rang again. They groaned loudly as Meghan's friend answered the phone impatiently. "It was my mom," Meghan said. "She just said that there was an accident, and I needed to come home."

We were in the living room of Meghan's home just up the road from her parents' house, a light-filled afternoon in the quiet weeks after harvest, as the year runs toward winter. She paused a moment, fighting back tears. She said she yelled at her mother, demanding to know what was going on, but there were no details. All anyone knew was that Dyrek had called Brent's mother Rita and told her that Brent was hurt. In shock, Meghan's friends helped her gather her things and then drove her back home, all the way from Lincoln. As they drove, Meghan kept hoping for the best—that this was a mistake or that Brent was hurt but not badly. "Usually, it's a casualty assistance crew that comes out and tells you, if it's really bad," she said, and no one had contacted Rita yet.

Once Meghan was home, her parents drove her to Clarks to wait with Rita for further news. Hours passed. As evening came on, two men in uniform pulled into Rita's driveway. "I guess that's when you know it's real," Meghan said. She wiped her eyes. "Normally, I'm not sad. Normally, I can talk about it." They soon learned that Brent's Humvee had been hit by a roadside bomb in Al-Qa'im, where the Sunni insurgency was now spiraling out of control. The improvised explosive device had killed Lance Corporal Zoucha and two other Marines.

Meghan said she would never forget the flood of emotion as the two casualty notification officers delivered the official news. "But the hardest day is the day the body comes home. You go to the airport and you expect them to walk off the plane, as good as the day they got home from boot camp. But they don't. It's nighttime and you go down to the tarmac and you receive the body." Dyrek was not hurt in the explosion, but he was detailed to return to Nebraska with his brother's remains. He was supposed to arrive on the same flight as Brent's coffin, but one of Dyrek's connecting flights was delayed. So Meghan met the funeral services crew and waited with Rita and Brent's family with the hearse. They loaded the casket and started the two-hour drive.

When they finally neared the funeral home in Central City, it was the middle of the night, but people lined either side of the highway. Those in cars and trucks fell into a convoy trailing the hearse. In the days after, Rita appointed Dave Beck, a family friend, to deal with the press. He told the local newspaper that Meghan was having a rough time accepting Brent's death. "They were the picture-perfect couple," he said, "the homecoming queen and the Marine. She was waiting for him to get home."

As the Zoucha family made arrangements, they soon learned that members of the Westboro Baptist Church of Topeka, Kansas, were threatening to show up at the funeral to stage a protest. A hate-filled family cult, all related to founder Fred Phelps, they believe that all tragedies are divine retribution for modern society's acceptance of Jews, Muslims, and gays. Since the wars in Afghanistan and Iraq had begun, they had been arriving uninvited at military funerals, chanting during eulogies and carrying signs that read "God Hates Fags" and "Thank You for Dead Soldiers." Zoucha family friend Carol Hanson was determined not to let the Phelps family disrupt Brent's funeral.

With Rita's permission, Carol began working with local law enforcement, the Department of Defense, and St. Peter's Catholic Church in Clarks to map out the funeral procession route. They figured out where protesters might set up and devised a plan for how best to shield family and friends. She called in the Patriot Guard Riders, a national organization of motorcycle enthusiasts created specifically to oppose Westboro Baptist, to form a line of flags to block the view of protesters. If their chanting got too loud, the guard was instructed not to engage them, not even to make eye contact, just to kick-start their motorcycles and idle the engines to mask the sound. In the end, the protesters never showed. Members of the cult were at funeral services in Beatrice, Nebraska, for Army Private First Class Ben Slaven of Plymouth, but a new state law had barred protesters from standing closer than 100 yards from a funeral parade route—and it appeared that the Westboro Baptist protesters were put off by their inability to disrupt services.

So on June 20 in Clarks, there was no protest, no clash, but the Patriot Riders and everyone in town had made sure that Brent would have what the local newspaper called "a red, white, and blue sendoff." Hundreds of people, all of them with American flags, lined the brick streets. "Stooped World War II vets struggled to hold large flags aloft," the newspaper reported. "Little Leaguers in uniform waved mini-flags and held hands over hearts; cheerleaders wore red, white and blue skirts. And many residents had run their front-yard flags only halfway up their poles, as they had every morning since the bad news traveled from faraway Al Bu Hardan, Iraq, to this town of about 375 people." The Patriot Riders circled through town and, with American flags mounted to their rear seats, led the procession to St. Peter's.

"It was crazy," Meghan said. "The whole town packed in there. It was wall-to-wall." She sat near the front, close to Brent's flag-draped coffin. Marine guards stood at each corner of the sanctuary. All told, more than 500 people, the entire population of Clarks plus more from far away, packed into the sanctuary, balcony, basement and then out onto the front lawn of the church. They listened as Father Ken Vavrina told them the young Marine had lived a life that honored the flag and died trying to protect it. "There are some people, and thank God, who say, 'What can I do to make life better for others?'" the priest said. "Lance Corporal Brent Zoucha was one of those young men." Following the services, about a dozen of Brent's fellow Marines carried the casket and led the procession out of the church.

The bells rang, and everyone poured out. The procession, led by veterans and Patriot Guard members, crept down West Amity, between the rows of flapping flags posted along the way. The line of motorcycles, trucks, and cars, flags mounted to the antennas and top racks and handlebars, inched down the crowded streets, past the fire station where the department had brought out the trucks and raised the ladders. Police blocked off Highway 30, allowing the procession to zigzag across the track, onto the blacktop, and then down the gravel road toward the country cemetery.

The wind whipped across the treeless green, the ropes of the flags snapping and making the poles ring. They said the Lord's Prayer. The Marines fired off a 21-gun salute. A lone bugle corps musician played "Taps." Two Marines folded the flag from Brent's coffin and gave it to Rita. Then Brent's coffin was lowered into the plot next to his father and everyone returned to their cars. The flags up and down the streets in town were lowered, folded, and put away. The COME HOME SAFE sign in Rita's yard was taken down and carried to the garage, replaced by a permanent memorial with Brent's official military portrait. When the semester started in Omaha, Meghan wasn't there. She decided to take the trip to Ireland that she had postponed, and waited until January 2007 to start classes. When she did finally go off to Omaha, away from home and on her own at last, she didn't find the sense of independence she had been looking forward to just months before. She felt lost and alone.

After her first semester, though, Meghan was determined not to go back to the farm for the summer. She took a job as a porter at an auto dealership in Omaha. "Just something to do," she said. "A way to keep busy." The shop manager thought she was great and kept offering to set her up with one of his former mechanics. Meghan kept putting him off, not sure she was ready and not wanting to explain. In August, just as classes were starting up again, Meghan confided to Rick that, more than a year after Brent's death, she had consented to a blind date with the mechanic. She wasn't sure yet what to think about dating him or anyone else, but all the girls in the office raved about how nice he was, and the guys in the shop praised his skill under the hood. Everyone agreed that he was a really hard worker.

Rick had been looking for help with that year's harvest, so Meghan put him in touch with Kyle.

IN THE spring of 1895, when Harry Harrington took over farming Thomas Barber's land, the family operation was already in serious trouble. Those 520 acres were considered some of the best ground in Nebraska, but severe droughts in 1893 and again in 1894, not only presented a hardship for the Harringtons but for the entire state and then the nation and eventually the world. Global failures of the wheat and corn harvests had posed a serious threat to the world's food supply, touching off famine in Europe, Russia, and China. Matters worsened when nervous stockholders and commodities speculators started to pull their money from the free-falling crop market and put their cash into reliable gold. So many investors followed this course of action, however, that it caused a run on gold—and set off additional waves of panic. The silver market collapsed, putting Colorado miners out of work and creating a tent encampment along the South Platte River outside Denver that was nearly half the size of the city itself. Attempts to slash wages in large industrial cities led to violent strikes and crackdowns among the Homestead steelworkers in Pennsylvania and the Pullman workers in Chicago.

Though remembered now as the Bank Panic of 1893, the crisis of the 1890s was actually a global economic depression that lasted almost to the end of the decade and was the worst downturn in American history until the deepest trough of the Great Depression. In Nebraska, it may actually have been worse than the Dust Bowl years. Charles H. Morrill, a well-known Nebraska farmer and banker who lived through those times, later recalled that farmers were forced to market their entire grain crops and even sell their hay. Without anything to feed their livestock, farmers starved their animals or simply butchered them for food. "The entire state," Morrill wrote in his memoir, "was almost in the grip of actual famine." And when the harvest failed again in 1895, the effect was devastating. In small towns across the state, the shortage of gold, which was used to back paper currency bank by bank, led ordinary citizens to withdraw their savings all at once. "Values were greatly reduced, merchants and banks failed," Morrill wrote. "Farmers could not pay interest on their mortgages; land could not be sold at any price; foreclosure of mortgages was the general order."

The situation was so dire that the denunciation of the gold standard by William Jennings Bryan at the Democratic convention in Chicago in July 1896, in what came to be known as the "Cross of Gold" speech, vaulted the former Nebraska congressman to national prominence and the top of both the Democratic and the Populist tickets for president that fall. Desperate farmers hoped that a western candidate would unseat entrenched, gold-hoarding powers; instead, Bryan lost to William McKinley, a "straddle bug" on the currency issue who was perceived to be in the pocket of moneyed East Coast elites and indifferent to farmers even as the harvest of 1896 marked yet another failure.

By then, the drought had grown so severe—even in water-rich parts of Nebraska, like York, Hamilton, and Polk counties—that trees a foot in diameter, planted as windbreaks after the locust plagues twenty years earlier, died all at once, and land prices fell to $15 per acre, even for properties with houses and barns. "Many farmers who were out of debt at the beginning of the dry years, and who had declared that 'no mortgage would ever be put on their land' were forced to mortgage to obtain food for their families," Morrill wrote. "Nearly every family discussed daily the question of the impossibility of remaining in Nebraska and debated where it would be wise to go."

Harry Harrington hung on through those years—but just barely. By 1902, the worst of the financial crisis was over, but the walls seem to have been crashing in on him. Harry and Lela now had four children: three daughters and a son, named Wayne. To hold onto Centennial Hill Farm, Harry sold the 260 acres he had inherited from his own father across the tracks from Benedict, located directly on the Kansas City &

Omaha Railroad with a grain elevator right at the field's edge. Some of the most desirable land in all the West, Harry sold the property for just $55 per acre. All these hard times and losses seem to have taken their toll. In those years, Harry is recorded to have admitted to an investigating police officer that he had purchased bootleg liquor from teenage brothers in Benedict and, in a separate incident, was brought before a police judge in Columbus for public intoxication and resisting arrest. Even as harvests slowly recovered and yields climbed, commodities prices dipped. It seemed that, whether it was a bumper crop or a bust year, the speculators profited and farmers like Harry suffered.

At the turn of the century, fed-up farmers who had long been loyal Republicans out of lingering gratitude to Abraham Lincoln for their land grants, now began flocking to the Populist Party and forming collectives such as the National Farmers Union. Under the banner of "Raise Less Corn, More Hell," they sought to create their own market power by controlling production during peak periods and sharing knowledge and labor to boost yields during hard times. Among those leading the Populist movement just twenty miles down the road from Harry Harrington was my own great-grandfather, L. C. Genoways. He started a successful corn bank, from which he sold his prize-winning seeds to neighbors to improve their output, and he won election on the Democratic ticket as the assessor of Hamilton County, vowing to fight on behalf of farmers at the capitol and governor's mansion in Lincoln.

In 1912, the State of Nebraska sought to boost sagging revenues by changing the way it taxed farmers, levying according to farm values as determined by the overall productivity of land in each county, rather than a flat fee according to acreage. L.C. gathered a delegation from central Nebraska and demanded an audience with the governor to protest. He spread out a map of the county on the governor's desk and showed him the thousands of acres along the Platte River, including the 160 acres where his own family lived, which he said were good for nothing but grazing livestock. He complained that, if the governor's planned tax increase were enacted, those acres would be valued and taxed at $78 per acre and similar land in York County would be taxed at $80 per acre—more than it could be sold for. But the state treasurer was unmoved. He asked L.C. at what rate the land in the area was returning across the board. "I do not know," L.C. replied at first to the huffing disapproval of the governor's staff. Finally, L.C. acknowledged that the central part of Nebraska, when taken as a whole, was returning 85 percent of its stated value. "You have the best farm land in the state," the governor replied angrily, "and you ought to be willing to admit it."

There are no surviving documents to explain why Thomas Barber finally sold his land in 1915 and moved away to California, leaving his daughter and her now six children without a land base, forcing the family to move to a rented farm near the town of Lushton. Maybe it was an inability to pay the stiff new taxes, or maybe it was conflict between Thomas and his son-in-law, or maybe it was just realizing how greatly farming had changed. Forty years earlier, Thomas had arrived with a horse-drawn mower and a threshing machine. Now, tractors were starting to replace horses, and hand labor was being replaced by gang plows, tandem disks, harrows, combines, and trucks. The new tractors made it possible for farmers to work large tracts of land, even if those plots were scattered across wider geographical areas, but they also required major capital investments, as much as $350, and were unreliable and expensive to fix.

Within a decade of buying the Barber farm, the new owner, Sherman R. Severn, had set a new record for speed of harvest in the state of Nebraska. Working only with his son, Severn was able to harvest 160 acres of wheat and oats in just eight days. With yields in York County rebounding from the drought years and nearly doubling the state average, the Severns had brought in 800 bushels of grain, delivering them by truck to the elevator. In Thomas Barber's day, it would have taken a team of men working sixteen-hour days nearly three weeks to match that feat. But buying new combines and tractors was still a risky business. In 1916, Wilmot Crozier, a farmer from Osceola in Polk County, bought a three-horsepower Ford 8-16 tractor, believing it had been manufactured by Henry Ford. Only after he discovered that the tractor didn't have nearly the power it advertised and would only pull a single plow did he discover that the Ford Motor Company manufactured Fordson tractors, not Ford tractors.

Given these rapid changes, it seems telling that Wayne Harrington, upon turning eighteen, left the farm in Lushton, taking apprentice jobs in mechanic shops in David City and Polk before eventually heading to western Nebraska, where he learned to repair tractors. In 1919, the year he returned to York County, Wilmot Crozier partnered with State Senator Charles Warner in drafting the Nebraska Tractor Test Law. From that point forward, any manufacturer wishing to sell in the state would have to supply one of its tractors to the University of Nebraska for performance testing. The law did wonders for farmers' willingness to invest in tractors. Sales soared, and Wayne opened a tractor repair garage in Lushton, where he worked on the new diesel-powered farm equipment. When his grandfather Thomas fell ill and eventually died in Long Beach in 1921, Wayne traveled to California and brought his body back to York County for a funeral at the Presbyterian Church. More than forty-five years after first arriving and buying land in Nebraska, Thomas Barber, in death, now owned no more ground in York County than a burial plot.

In 1928, Wayne married Eunice Kniss, a native of nearby Sutton, who had attended the University of Nebraska for a year, earning her teaching degree, before traveling to various one-room schoolhouses, finally arriving in Lushton. Wayne's repair garage, by then, had grown into an implement and hardware dealership, a grocery store, a livery stable, and a Ford car agency. In 1933, Eunice gave birth to the couple's only child, a son named Thomas. When Tom was seven, the family moved to Fairmont, at the intersection of Highways 81 and 6, nicknamed "The Crossroads of the World," to have better access to major roadways. Along the highway, Wayne expanded his operations to include a lumberyard, a grain elevator, a motel, a café, another auto dealership and auto repair shop, and a successful trucking company bringing salt from the underground mines of Kansas to cattle ranchers in Nebraska.

For once, the family seemed to be thriving through hard times. While the rest of the country was gripped by the Great Depression and World War II, Wayne's business continued to grow and prosper. If Wayne was a severe father and strict disciplinarian, Tom nevertheless grew up in a boy's wonderland, drinking malts at the soda counter, hearing stories of the world from long-haul truckers, and learning all about heavy equipment from hard-drinking and chain-smoking mechanics. Among the farm kids of the Lushton school district, he was seen as rich. When his classmates voted to take a field trip to Lincoln in spring 1946, Wayne's trucking company provided all of the buses, ferrying the children from the state capitol building to a tour of the city to the 4-H building of the State Fairgrounds for a matinee performance of the Shrine Circus, all buses emblazoned with custom paint jobs reading WAYNE W. HARRINGTON, TRUCKING.

On May 21, 1949, Wayne's father Harry died in Fairmont, but that same year, Wayne had the opportunity to buy back the Centennial Hill Farm. To make it work, Wayne would have to sell all of his existing businesses. Eunice later remembered that he decided to do so "mostly out of sentiment for the old place," but Jenni told me that she thought that Wayne felt pressured by his mother to reclaim what her father had lost decades before. Whatever the reason, Wayne entered into farming with a business-like attitude—acquiring not only the original farmstead but as much adjacent land as possible, picking up low-lying properties that had been used as basin land. From his work as an implement dealer, Wayne knew that companies like John Deere and International Harvester, which had grown by leaps and bounds on government contracts during the war, were now turning their attention to developing heavy equipment for farmwork. In particular, they had already developed pump systems capable of siphoning water from rain catchment ditches at the edges of farm fields, in order to use it for sprinkler irrigation.

Rather than have Tom help him on the farm, Wayne sent his son away to Shattuck Military Academy in Minnesota. "He was really hard on my dad," Jenni told me. "You just feel like there wasn't much of a connection between dad and son." She said that once, many years later, she asked her father if he thought that Wayne had loved him. He thought about it carefully, she said, and finally answered that he thought that his father had loved having a son who was such an accomplished athlete. But even that wasn't simple; his father had encouraged him to go out for football, but Tom was tall and lanky and excelled at tennis and golf. He had dreams of skiing, which was not at all what his father had in mind. "You can't say that he loved you?" Jenni pressed, but she said her father refused to answer.

With Tom now out of the house, Wayne went to Germany to a displaced persons camp and brought back a Latvian refugee named Arnold Jaunzemis and his family to run the farm. Wayne also caught wind of researchers from the University of Nebraska's Department of Agricultural Engineering, led by Professor John F. Schrunk, who were digging test wells across the state in search of groundwater that was close enough to the surface that it could be pumped into catchments and used for irrigation. They discovered that the water table in eastern Nebraska was not only shallow but also that the Ogallala Aquifer in that area rapidly recharged with rains from the northern Sandhills and snowmelt from as far away as the Rockies.

To sell area farmers on the idea of digging wells and installing pumps, Schrunk organized an "irrigation tour" of the most successful wells around York, culminating with a visit to the Centennial Hill Farm, where Wayne was then raising 250 acres of irrigated corn. "He uses two wells," the *Nebraska State Journal* reported, "one pumped with a caterpillar diesel and one with propane. The one pumped by diesel puts out 2,000 gallons per minute." After the well demonstration, the farmers gathered for "a ton and a half of watermelon and several hundred bottles of pop," all enjoyed in the shade of Wayne's new 15,000-bushel corn crib and 6,000-bushel small grain crib. He also showed off his newly constructed air-conditioned hog barn and a brand-new corn picker he had purchased for the upcoming harvest.

"Wayne was an innovator," Rick told me one day as we drove from one property to another. "But, you know, it wasn't too long before it cost him."

KYLE HOPPED out of the Kawasaki Mule and hunched against the blowing snow. It was almost dark, and the clouds seemed to be gathering force, sending waves of heavy flakes swirling through the four-wheeler's headlights. After a mostly dry early winter, central Nebraska was seeing its first real storm, and the wet snow was filling the furrows and covering corn stubble. Along the edges of one field where the cattle had been turned out to feed and then cordoned off with an electric fence, the prairie grass was so heavy with snow that it had lain down across the electric wire, tripping the circuit breaker. Kyle had come out to find the source of the problem and make sure none of the cows were out. Before driving the perimeter, he checked first to make sure that the solar panel and the car battery that ran the system were properly hooked up. It didn't look like anything had been disconnected by snow or wind. Kyle slid one insulator down the rebar fence post until the electric wire was flush with the ground and climbed back behind the wheel of the Mule. He cleared the melting flakes from his glasses. "It is really starting to blow," he said.

As Kyle drove the edge of the acreage, the darkness deepened until his headlights, bright with swirling snow, were the only light. Every now and then, he would stop the Mule, jump out again into the wind, bend back the broken stalks of prairie grass sagging onto the wire, and then slide back into the Mule. As we drove along one row of furrows, the bodies of the Angus cattle emerged faintly, the black of their snow-sifted bodies against the blacker sky. Some had white foreheads and noses. Mottled and pocked with patches of black, their faces reflected the beams of the headlights like the cratered face of the moon. When Kyle would step out into their midst, they would startle back into the darkness, watching warily as he checked the fence. At last, when he thought that every inch of the wire was cleared, Kyle drove back to the control box and reset the system. It instantly began to tick, the telltale sound of the pulsing electric fence. He smiled broadly: "Well, at least that's one thing taken care of."

Kyle was born in Mason City, Iowa, and adopted at birth. He lived in Indianola, Iowa, until he was four years old and then moved to Omaha. Kyle never knew his birth parents, and the parents who raised him never talked about them. That mystery became a kind of running joke with his friends. One, who was Hispanic, always kidded that Kyle's love of Cholula sauce and tacos just proved that he was really Mexican. Another joked that he was too tall and beefy to be a Mexican. It was all meant in fun but left Kyle wondering, and it's hard not to read his shyness, his quiet laugh, and gentle demeanor as a low-key way of trying to fit in. Kyle is so agreeable, so quick to pitch in, that Meghan's description of his hot temper is almost impossible to imagine. "Oh, he's just fooling you," she said.

We were back at their house now, drinking cups of blazing hot coffee while the snow continued to fall and build. "Yeah, the very first stories I ever heard about him," Meghan said, "before I even met him, were about what a hothead he was." Kyle rolled his eyes and shook his head. He rubbed his legs, still warming up from the cold. "That was kind of a unique situation," he said.

After his sophomore year of high school in Omaha, Kyle took a job at an auto dealership, just emptying trashcans and sweeping up at first. But the manager soon figured out that Kyle was good with his hands and started training him to be a mechanic. He showed such skill that his supervisor agreed to sponsor Kyle for the Chrysler-supported program in auto repair. If accepted, the dealership would pay half his tuition to the branch of Southeast Community College in Milford, just outside Omaha. He would get an associate degree of applied sciences, and if he made the dean's list each semester, he would come back to the shop with a raise. Kyle made the dean's list each of his first three semesters but narrowly missed it in his final semester before graduation.

"So they didn't give me the raise," Kyle said. But what really bugged him, Kyle said, was that without the promotion, he didn't have the standing to pick his repair jobs. Instead, the service writers would dictate which mechanics were assigned to particular jobs—who got the repairs that would pay well and be easy to complete, who got complicated repairs that would take time and expertise to complete but didn't involve expensive parts so they made it hard to earn commissions or build toward bonuses. "They were giving me shit-work," Kyle said. "And whenever I complained, the service manager kept telling me, 'The other guys have families, and they need to support their families.' And I was just like, 'I got to fucking eat too, you know.'" Finally, one day in a fit of anger, Kyle threw a wrench through the service window, shattering the glass.

After that, Kyle wanted to get as far away as possible. He moved five hours west to Ogallala. A friend from Southeast was from there and helped Kyle land a job in the shop at Schmidt Motors. Not long after, Kyle's old shop foreman Barry in Omaha called him up and said that there was this new girl working at the dealership. Kyle had to meet her. "No way," Kyle said, adamant that he didn't want to be set up. "I hate that shit." The girl, of course, was Meghan, who had taken the job to stay busy after completing her first semester of college. Barry thought that Meghan would be perfect for Kyle. What he didn't know was that Meghan's last boyfriend had been killed in Iraq, and she was still grieving. Plus, the first story she had heard about Kyle was about how he'd smashed up the service department on his way out from quitting. "I was just like, 'I do *not* want to meet up with this guy,'" Meghan laughed.

But after Kyle moved back to Omaha, Barry invited the two to a group lunch at a sports bar, not telling either that they were being set up, just hoping they would hit it off. Meghan said when she heard Kyle was there, she steered clear of him. "I didn't even fucking talk to him," she said. More months passed. In the meantime, Meghan said that every mechanic in the shop was telling her how great Kyle was, even as they were also telling her more about that fiery temper and a tendency toward taking big risks. (A favorite story involved a dirt bike accident and Kyle having to be restrained by hospital staff while they stitched him up.) Finally, Barry couldn't take it anymore.

Meghan had come into a pair of tickets for the River City Roundup, a giant rodeo held every year in Omaha, but she didn't have anyone to go with her. She made the mistake of mentioning this to a coworker, who passed the word along to Barry. Barry took Meghan outside of the dealership. He looked her in the eye and said, "You're going to invite Kyle to the rodeo." He dialed Kyle's phone number and gave Meghan the phone. When Kyle picked up, Meghan was so flustered that she just blurted out the invitation. "Hey, do you want to go to a rodeo?" she asked.

Meghan turned to me, her eyes wide with emphasis. "Dead fucking silence," she said. "Just nothing from the other end. Like, a *really* long, awkward pause."

Kyle said that he knew Barry was behind this, that he'd been working for months now to set him up with Meghan, and had been pushing so hard that Kyle was resisting out of sheer stubbornness. But now, put on the spot, he couldn't think of an excuse. He mustered just a single word: "Sure." Meghan howled now with laughter. "Real romantic, right?"

But they went to the rodeo together, and they had a good time. After a few beers, they loosened up and started talking. At some point, Kyle mentioned that he was still looking for work, and Meghan said that her dad needed someone to help with harvest on their farm. Kyle talked to Rick and soon moved into the little white house on Centennial Hill. "I was just working for her dad, and Meghan was in Omaha. She wasn't even out here." At the end of harvest, Kyle got another temporary job, driving a plow and repairing trucks for a snow removal company in Omaha. A couple of times, Meghan rode along as Kyle was plowing driveways and private roads during snowstorms. But he said they were still more friends than anything else.

"When March came around, I needed to find a permanent job, so I applied at Caterpillar." The company hired him to work as a repairman at the dealership outside Omaha, charged with fixing breakdowns. It took almost a year, seeing each other for dinners and movies after his shifts at the dealership, before the two finally admitted to themselves that they were dating. When Meghan moved back to the farm after graduating from Metro, she said it only made sense for Kyle to come with her.

⁕

OF ALL the stories of Centennial Hill Farm, one stands out—a passed-down tale of super-human grit but with so few concrete details as to create a certain shroud of mystery. It was September 1951. Wayne Harrington was just beginning the corn harvest. Maybe he'd decided to get the corn out early. The weather was perfect, sitting squarely in the 70s with no rain for the whole month. Maybe he had hoped to get a jump on the market but then found the stalks still a little green. Those parts of the story are lost. But whatever the reason, he had barely begun on the field when the picker jammed. Wayne climbed down from the harvester and cleared away the mouth of the machine, but the jam was deeper inside. He reached in, dislodging the problem cob, and the teeth of the picker grabbed his fingers, his whole hand up to the wrist. He was mangled by the machine and caught. Worse than that, he was alone in the field and pinned. "He had a *booming* voice," Jenni told me. He called out for help and was finally heard by the women from the Latvian family he had brought back from Europe two years before.

The women came running from the farmhouse. Following Wayne's instructions, they retrieved his tools from the barn and took apart the gears of the picker. Wayne freed himself from the teeth of the cobber, wrapping his arm—with what, no one knows, his shirt or his coat—and then drove himself more than thirty miles to the hospital in York. But there was too much damage to save his hand. Two days later, doctors were forced to amputate. "People say it was really hard on him," Jenni told me, "and that he was never the same. It took a lot out of him."

That same fall, hoping to please his father, Tom enrolled in the University of Nebraska's school of agriculture and went out for the Nebraska football team. Tom later told his kids how he'd gotten beat to hell trying to prove himself to the coaches but still failed to make the squad. There would be no trips to Memorial Stadium to boost Wayne's spirits, no gridiron magic to take his mind off his loss. Tom did make the university tennis team, but his father never traveled to Lincoln to see any of his matches. Just a few months later, in February 1952, Wayne was out in a winter storm, delivering feed to his cattle, when his truck skidded off of icy Highway 81 and crashed into a tree. He suffered a fractured skull, a crushed rib cage, and multiple breaks in both legs. He held on for a few days at the hospital in York but finally died from his injuries.

"I've always wondered," Jenni said. "Did he lose control

of the truck, because he only had one hand? Or did he—"
She didn't finish her thought. But then I wondered, did Jenni
question whether Wayne's crash had been an accident? Was
that buried somewhere in the subtext of the repeated tellings
of the story? She didn't go further. "I don't know," she said.
"But that's when my dad came back to run the farm."

Tom had completed just one semester of ag training, had
never lived on the farm, and didn't have any practical farming experience. His mother took over running the books
and managed all aspects of the business. Even after Tom had
married his wife Karen and had four little girls, Jenni said
that she remembered how her grandmother would spread her
papers out across the kitchen table to make sure every line
balanced. "We just thought, *wow, she's the queen of the family*—
and she *was*." Some of her grandmother Eunice's attention
to detail arose from Tom's freewheeling approach to farming. Because he wasn't bound by any tradition, he was more
forward-looking than many of his neighbors, but he was also
taking risks and often guessing wrong. "He always liked to
live on the edge," Jenni told me. "So he always had cattle deals
going south or found himself in the wrong position on the
commodities market."

But Wayne had bought the family land back at a lucky
moment. In the 1950s, the irrigation projects Wayne had
helped promote suddenly took off. For the first time, groundwater could be delivered to the fields by "flood irrigation," a
method of pumping into containment ponds and then releasing water through a system of wide, open furrows. And soon,
Tom and others began installing the first irrigation sprin-
klers. The area would never again be at the mercy of droughts
like they had seen in the 1890s and 1930s. When center-pivot

irrigation was developed just up the road in Columbus, Tom was one of the first to buy in, so that, when blistering temperatures struck Nebraska in 1953, Tom's farm, with its well-irrigated fields, saw record yields, and Centennial Hill quickly became some of the most valuable farm ground in the world.

With the help of new chemical fertilizers and pesticides, high-quality hybrids, and his ever-expanding irrigation system, Tom was soon appearing in local newspaper ads announcing top yields for the county. At the same time, his family was growing and prospering. Tom and Karen had four daughters—Abbi, Terri, Jenni, and Heidi. And he set about using his newfound wealth to build a modern farmhouse for the six of them.

Tom hired Dewey Dearing, a well-known architectural firm in Colorado Springs, to build the new house, patterned after the ski lodges that Tom admired in the Rockies. Everything about the place was eccentric. Tom requested a sofa suspended from the high ceiling by a heavy chain, so it would swing. He had the floors made of stone and covered the walls with thick carpet, so dense that guests had a hard time finding the light switches. "You ought to see them try turning on the lights," Tom told the *Omaha World-Herald* when they sent out a reporter to write a story about this unusual farmhouse. Tom even personally designed built-in bookcases with slanted shelves on the idea that this would eliminate the need for bookends. Of course, it damaged books and made those at the bottom of the incline almost impossible to budge from their place. "I goofed," Tom told the reporter with a shrug.

Jenni winced, remembering the photograph that accompanied the article—a picture of her, about age four, next to her mother in the swinging sofa. The picture was reproduced using the new Colorphoto process, which allowed four-color reproduction but with a rather limited palette. "It was an orange sweater with a plaid skirt—that kind of matched the couch," Jenni said. "I just about disappeared!" This was the dichotomy of their childhood: daily chores tending the chickens and cattle, with evenings spent in a house that stretched the limits of sixties style. But Jenni said it never seemed incongruous to her or her sisters; they knew that their neighbors considered them oddballs, but she said that all of the girls identified with their father and his progressive mother.

Jenni's sister Abbi remembered going to visit their grandmother Eunice at her new home in Omaha, where she had gotten deep into social activism through the Methodist church. "She always loaded us in the car, and we would venture into the city to see firsthand what she had been working on. We visited a black radio station she helped start. We would drive out to Boy's Town to see what they had going on." Their relationship with their father was more difficult, especially as they grew older and more rebellious. When he made them drive the tractor at planting or harvest, they would don their bathing suits to work on their tans, just in protest. "We drove him nuts a lot of the time," Jenni said, but she also said that her father often seemed to be looking for a fight.

Once when the girls were in high school and the state was in the midst of a drought that was devastating a neighbor's milo crop, Tom dug a culvert across the county road and laid a pipe to divert water from his catchment pond to the parched field. When someone complained, the county sheriff ordered the pipe removed. As soon as the crew was gone, Tom put it back. The county responded by issuing a restraining order. Tom, in turn, submitted a formal request for a temporary permit "to cross their goddamn road." The York County Board of Commissioners met to consider the matter and ultimately denied the request 3 to 2. "Tom Harrington's a pretty good fellow," one of the commissioners said after the meeting. "You can't help but like him in many ways." But, he added with a smirk, "he's rather headstrong." Tom responded by threatening to file a lawsuit. He conceded that he probably had no case but told the *Lincoln Evening Journal* that he was in search of an attorney who was willing to sue the county "for damage to the milo and for being stupid."

A few years later, Tom attempted to organize a county-wide holdout of property tax payments in protest of a state tax reappraisal, which set the value of top irrigated land in York County about 20 percent higher than values in surrounding counties. "There are a lot of people who are taking it in the shorts over property taxes," Harrington told the *Omaha World-Herald*. "Most farmers will be paying $10 or $12 per acre in taxes on their irrigated land." About fifty landowners flirted with Tom's idea of an organized tax boycott, but the possibility finally fell through over concerns that a lack of tax dollars might shutter public schools in the county or even force the seizure of property in order to collect those payments. Tom, thinking they were all cowards, vowed never to talk to any of these neighbors ever again.

Jenni said that the contentious nature of the household drove her oldest sister Terri away to college and then to Denver, where she became a lawyer. But the other sisters, though they also went away for school, returned after Tom announced that he planned to retire and leave the farm to the four of them in a joint trust. The land was divided equally among them but still run as a single operation, requiring all decisions to be reached by consensus. That closeness led to years of fighting and acrimony and eventually to stony silence, even as each of the four sisters refused to sell her shares. Instead, lawyers and negotiators were often called in to broker even relatively routine business decisions, and no one would budge. Divided by what they shared, the sisters remained—and still remain—bound to Centennial Hill Farm, the center of the Harrington universe, for better or for worse.

Nearly a decade of tension followed until, unexpectedly, a threat arrived in the form of a land agent working on a planned pipeline project. He said that he represented a foreign corporation and that he was there to secure an easement—or they would face having their land taken by eminent domain.

RICK REVVED the engine of his four-wheeler, sliding through the prairie grass. In only a few years of letting the land lie fallow—just a spot to graze his small herd of black Angus for a few months out of the year—head-high stands of big bluestem and Indiangrass had sprung up. He chugged to the top of a bluff and cut the engine. It was quiet there; the only sounds: mooing cows on a neighbor property to the north and the sighing of wind whipping across the meadow. Rick's own cattle chomped drowsily, their loose jaws churning and ears twitching as they watched us from afar. Now February, the last drifts of snow hid in the lees of ridges and cedar trees. Calves would be coming soon, and Rick was hoping to fatten the expectant mothers before they gave birth and started nursing. They rested, dark and still, seen only through the shifting veil of ochre grass.

Otherwise the fields appeared continuous and empty. Clouds cast shadows that seemed to move like galleons across the green and yellow waves. "Prairie ships," Rick said. Farther in the distance, the wide bend of the Platte River curved from Central City, with its grain elevators and ethanol towers, around to a grove of trees to our east, where a Pawnee camp had been for hundreds of years. My own grandfather talked about his father going to that camp to trade in the 1880s. It was this view that Rick wanted me to see, and he pointed out a cluster of depressions along the edge of the bluff—a group of Pawnee graves, he believed. He could understand why they would have buried their dead here, with this panoramic view of the river valley.

"I come up here sometimes to get a little perspective," Rick said. "It's easy to get lost in the worries of the day-to-day. You worry about money, mostly—yields or prices—and you can forget to appreciate what you have." Rick acquired this land, about fifteen miles northwest of Centennial Hill, when Meghan and her siblings were still little. Right from the beginning, he envisioned this as the view from his dream house—a place of his own, close to Heidi's inherited acres but with an even deeper history and one undefined by her family. "When we bought this land, it was a sealed bid," Rick said. "When we found out we'd won, Heidi was jumping up and down. But I bent over like somebody had kicked me in the gut, because I knew that for the next ten years I had to produce every year and as efficiently as possible. It would be a decade before we could even think about re-mortgaging everything to where we could build."

In 2001, when all of the kids were coming into middle and high school and wanting more room for themselves, Rick decided the time had finally come. He hired an architect to help him realize his vision of a giant timber frame home, open from the ground level to the rafters, with huge bay windows and a wraparound porch to give sweeping views of this land overlooking the river. But he said that between the cost of the architect, hiring expert builders, and high-end materials, he couldn't afford a general contractor, so he oversaw all the construction himself. One day, as the timber framers from Missouri were nearing the end of their work—the tall beams raised and walls blocked in, the plywood subflooring installed and everything but the stairwell completed—Rick climbed a ladder, stretching from the basement to the ground floor, to check out the progress of the finishes.

Rick interrupted his story to walk around to the back of his four-wheeler. He leaned against the seat, looking up at the house on the hill. The field of grass in between switched back and forth in the wind.

"I was all the way at the top of that ladder," Rick said, pointing toward the peak of the roof in the distance, "and just going down, when one of the guys below called up to ask me a question. I turned and that wiggled the ladder enough that the bottom shot out." As the feet of the ladder skidded across the concrete floor, the top slid down the supporting beam and pitched Rick off into the air. "I hit my head on the first floor going down," he said. "That's the last I remember. They found me in the basement with my head hung through the bottom rung of the ladder. They called Heidi and said, 'You better get over here. We don't know if Rick's going to make it.'"

Meghan told me later that she was with her mom and siblings, moving cows in the neighboring field, when the call came. By the time they arrived, the Hordville ambulance was already on the scene, its single light on top still lazily turning as the medics gave Rick mouth-to-mouth. They revived him and took him to the Central City hospital, where he was diagnosed with a concussion, a dislocated shoulder, and several broken ribs. But they didn't find any internal injuries, so they told him, "If you can shower, you can go home." Rick had Heidi help him into the bathroom and under the showerhead, and then she slid his shirt over his shoulders and buttoned his pants. He was determined to keep his promise to hold a cookout for the timber framers before they finished their work and had to go home to Missouri. But a week later, still in pain, Rick gave in to Heidi's insistence that he see a doctor in Lincoln, who found that he had undiagnosed fractures in both of his wrists and more broken ribs. Rick gave me a sideways smile. "That house almost killed me," he said, "physically and financially."

Rick was nearly out from under that burden when he received a strange letter in 2010. An oil company was writing to inform him of an upcoming town hall meeting at the church in Hordville. The company, TransCanada, was planning to build a pipeline, stretching from the tar sands mines in northern Alberta to refineries on the Texas Gulf Coast, and they wanted to meet with area landowners. Rick soon learned that the proposed route would cut directly through this piece of land that he had almost died for, trying to turn it into the perfect place to retire with Heidi. He went to that meeting feeling defiant. He had two natural gas pipelines under his grazing land in Curtis and knew firsthand what kind of impact digging trenches can have. "The land is never the same after it's disrupted like that," he said. "I don't care what they say. It's never the same." But a TransCanada land agent from Tennessee, who was also an ordained Baptist minister, told everyone that the company wanted to do right by the community, that he could see that they were the salt of the Earth and that he would make sure they got a fair deal.

Rick was dubious—and the more he researched the pipeline project, the less he liked it. The refining of tar sands crude is among the dirtiest fossil fuel processes in the world, contributing to climate change in ways that, in Rick's view, were a serious threat to Meghan's future on the farm. More than that, he couldn't find proof that any of the heavy diesel that would come out of the refining would even be legal to burn in the United States, and the refining byproducts, like petroleum coke, were expected to be burned in China, both undercutting American coal and supplanting it with an even more toxic substitute. Most worrisome of all, that very summer a similar tar sands pipeline owned by another Canadian oil company ruptured in Marshall, Michigan, severely contaminating nearly fifty miles of the Kalamazoo River with oil, toxic diluents used to make the heavy crude move through the pipeline, and still more chemicals used to break up and absorb the slick.

"They asked, 'Oh, don't you want to get your fuel from a friendly neighbor?'" Rick said. "But they never talked about all of the refined petroleum products going overseas or about how they were major greenhouse gas emitters. I mean, what the fuck? Why would I want a pipeline pushing toxic sludge through my soil, through the Ogallala Aquifer that we depend on for irrigation and drinking water, so some foreign oil company can make a buck? How is that being a friendly neighbor?" But Rick confessed that for all his anger, he was also afraid.

Along with the promises of a fair price and a fair deal were oblique threats of land seizures if farmers and ranchers dared to resist. He stalled on giving TransCanada a definitive answer.

"I was polite, but I was dragging my feet," Rick said. "After it became obvious that I was not going to sign, they threatened me twice with eminent domain, on the phone and once in writing." After he received that letter in July 2010, Rick gave in and signed the easement. He regretted the decision even before he put the documents in the mail, but he felt he had no choice. TransCanada was a $50 billion multinational, and he was a self-described "little guy," one man against a fleet of lawyers with nothing more on his side than a gut feeling that what they were doing shouldn't be legal. What could he do? "They crushed me," he said with a shrug.

Six months later, TransCanada contacted Rick to tell him they'd changed the proposed route, skirting around his property after an early cultural impact study suggested the likely presence of ancient Pawnee artifacts on his land, perhaps even graves, exactly as he had always suspected. Rather than deal with the red tape if one of the company's backhoes hit human remains, TransCanada had decided to push the route to the east. At first, Rick was relieved. "I said, 'Okay, if I give you the money back, will you give me my signature back?' They said, 'No.'" Worse still, when TransCanada revealed the new planned route, Rick discovered that it crossed a section of his sister-in-law Terri's land, just west of Centennial Hill— one of the properties where he raises seed corn for Pioneer, the cornerstone on which much of his operation now rests. Though the pipeline was no longer passing by the doorstep of his dream home, the new route threatened to be even more perilous to the family business.

Rick and Meghan convinced all of Heidi's sisters to join up with two antipipeline organizations, the Nebraska Easement Action Team (N.E.A.T.) and Bold Nebraska, and they even started their own local group in York County called the Good Life Alliance. "They may have crushed me as an individual," Rick said. "But there is strength in numbers and all of us standing up for the right thing." They started going to meetings and speaking out, writing editorials and placing full-page anti-pipeline ads in the York newspaper. Rick told me that, if there was any silver lining in living with threats from TransCanada, it had been watching Meghan blossom into one of the most vocal and articulate young leaders of the pipeline resistance.

Meghan and her siblings had been outspoken liberals since they were in high school. "During the Kerry-Bush election we had a vote in our political science class," Meghan said later, "and my twin sister and I were the only two that voted for Kerry. Our teacher, said, 'Oh, you Hammond girls, you just want to be different.'" But now, with their land under threat, some of those same neighbors were pleased to see Meghan speak up at zoning meetings of the county commissioners and committee meetings at the Nebraska legislature. She combed over the government reports and prepared careful statements. When the U.S. State Department's own study estimated that the project would create only thirty-five permanent jobs along more than a thousand miles of pipeline in the United States, Meghan was quick to seize on the point. In May 2013, she rose at a massive public hearing at the fairgrounds in Grand Island, Nebraska, held by State Department officials as part of the agency's approval review process. Still wrapped in a scarf from the freak spring snow swirling outside, she leaned confidently into the mic. "How can you risk our land and water for thirty-five permanent jobs?" she challenged. "If you want thousands of jobs, you will find it in sustainable energy."

It was that statement that gave Jane Kleeb at Bold Nebraska the idea of building a barn, powered by solar panels and a small wind turbine, on Terri's property. "It was kind of a 'put your money where your mouth is' kind of proposition," Meghan said. "We kept talking about wind and solar, but now we had to show that we believed in it enough to actually build it." Within a few months, Kyle and two of Meghan's cousins led construction, as dozens of volunteers from around the state helped to raise the structure in just four days. Billionaire environmentalist Tom Steyer flew in from San Francisco for the ribbon cutting and put up money for a viral video about Meghan and the barn. Heidi and her sisters posed together with Steyer, all smiles. But it wasn't long before the neighbor just to the south wrote to Rick to withdraw from the contract allowing the Hammonds to farm two quarters of his ground. Meghan told me later that she felt transformed by that betrayal.

"Losing two quarters of ground, that was huge—*huge*. And it's changed how we feel forever," she told me. "We wouldn't change what we've done with the pipeline, but our neighbors are our neighbors forever, you know? For generations and generations and generations. And now we have neighbors who won't even wave to us, so how are we supposed to trust them now?" Most of all, she said she was worried about the impact it had had on Rick. He dropped from view for months during that time. Some days he didn't leave the house, even during harvest. He regretted having signed the agreement with TransCanada, and he regretted that having fought back had brought unintended consequences. "He carries a lot on his shoulders, because he feels responsible for all of us," Meghan said.

At times like that, the weight can seem to grow with each piece of bad news. In November, the Republicans won their largest majorities in the U.S. Senate and House since the backlash against the Democrats after the stock market crash in 1929. New majority leaders vowed that final approval of Keystone XL would be their first item of business. In January, despite President Obama's veto threat, the House of Representatives approved the Keystone XL Pipeline Act, and the legislation passed the Senate on a 62-to-36 vote—still five shy of the 67 needed to override a presidential veto but with intense horse-trading expected in the effort to muster the remaining votes. Even if the Democratic Caucus could sustain the veto, talk was spreading that Obama might approve the pipeline as part of a larger deal toward sustainable energy.

Rick worried that the legacy he had spent a lifetime building was about to become a bargaining chip between politicians in Washington and corporate interests with holdings around the world. Was there anything he could have done differently? At times like these, he worries not just about the hardship of the moment but that he may have made a fatal error, the mistake that leads to future struggles for Meghan and Kyle or, in his worst nightmares, to losing everything and having to hold a farm sale. "He wants to make it better than the way him and my mom started out, but we don't need a bed of roses," Meghan told me. "They struggled. We will struggle. There's no way around it, but he always tries to make our lives as good as possible at his own expense. He's very hard on himself—*very*."

I couldn't help but hear Meghan's words in my head as Rick leaned against his four-wheeler, looking out over everything he had assembled and built and tried to defend. "We're extremely lucky," he said. "Almost everything you see is ours—from the lower farm to this bluff to the house up on the hill. And that's a blessing." But as soon as the words were out of Rick's mouth, the old worries seem to creep back in. "When I'm gone," he added, "I'll be able to leave each of my four kids at least as much land as Heidi's dad left us. That's the hope. That's what all the work is for."

BRANDING CALVES

April 2015

On a sun-soaked day in mid-April, the Hammonds corralled the cattle back into the paddock behind the old barn. The nursing mothers were separated from their spring calves, just as they had been in the fall, and then locked into neighboring pens for the night. Rick said it was probably eleven o'clock before the cows settled down and stopped mooing, slipping into sleep in the darkness, but they were up again before dawn, calling to their calves. By mid-morning, when the veterinarian arrived, the cows were noisy and rattling the gates of the corral. The sun was pale-bright again, washing everything into pastel hues, thin wisps of high cirrus clouds the only interruptions on the cornflower sky. The pristine clarity and spring-like promise in the air, even touched as it was with the last of winter's chill, seemed strangely at odds with the task ahead.

It was branding day, one of Meghan's least-favorite parts of the whole farm year, and she was feeling short-tempered and stand-offish. She wore a down vest over her sweatshirt and a loose-knit stocking cap, with her long hair swept to one side. When her temper flared, she would rip off the cap and tuck it impatiently into her back pocket, then pull it back over her ears when her mood cooled. Everything about the day seemed to be getting on Meghan's nerves. She didn't like how stressed the calves were: mewling and skittish, huddling together, face-in, in the corner of their pen. Seeing how jittery and strained they already were, she was pissed that Dave was using an electric prod to hustle some of the calves down the chute, and she was surly about the very idea of hot-branding. "It is hard on the animal," Rick conceded, "but they bounce right back." Meghan just didn't see any good reason for it. In an era when consumers are concerned about animal welfare, she worried it was a potential deal-breaker with boutique buyers.

More than that, it was unnecessary work, considering that there are more modern and pain-free methods of keeping track of individual animals. The cattle already had ear tabs to identify them within the herd. Those cheap plastic tags were hand-numbered by Kyle and catalogued with corresponding health information in Meghan's record books, an old-fashioned way of keeping inventory of the herd that required her to mark down information on the animals one by one. When the first calf lurched into the cradle, the trap closing around it with a hollow clang, Meghan yelled out, "What's the number?" Her older brother Jesse, back in town for the time being and charged with applying the brand, called back, "That's one-twenty-two." She echoed, "One-two-two," entering the number into her spiral notebook. Then she wheeled away as Jesse lifted the brander.

If, instead, the Hammonds used electronic identification (EID) ear tags, which have a fifteen-digit visual label specific to each individual and also contain a chip that can be read by an EID reader, then there would be no need to brand the cattle at all. The high-tech tags would make it easier to monitor animals both within the herd and to track down strays in the event of an escape. When we talked about it later, Rick allowed that EID tags would work fine for maverick calves or cows that wander off when an electric fence shorts out, but those ear tags are easily removed by anyone whose intention is theft. "You don't want your herd to cross the fence, and EID won't cure that," he said. "Especially if someone rustles them."

The idea of modern-day rustling sounds incredible, but with the price of cattle reaching record highs, worries about calf thefts had ranchers on edge from Texas to the Dakotas. The Nebraska Brand Committee, tasked with overseeing cattle brand registrations and enforcement over the western two-thirds of the state, reported that it had recovered more than 1,500 lost or rustled cattle between summer 2013 and 2015, at a total value of more than $5 million. Only a few weeks before the Hammonds starting their branding, three Omaha men had been arrested in connection with a series of at least seventeen thefts from feedlots and farms from western Iowa to Lincoln.

While that case was unusual in its daring, it was hardly isolated. The newspapers had been crowded with rustling stories, often simple crimes of impulse or opportunity, a weak moment when an unbranded weaned calf was spotted alone

Meghan Hammond closes up newborn chicks
in the henhouse for the night.

Rick Hammond stands in the stippled light of his family barn.

Kyle Galloway steers the John Deere harvester by hand
across a neighbor's field of soybeans.

Meghan hands dinner off to Dave Tyson. Corn harvest runs
late into the night, and meals are often eaten at the wheel.

Meghan reads while a load of soybeans fills a grain bin for later sale.

Jenni holds a photograph of her grandfather checking out his
new Farmall tractor, shortly after he bought back
Centennial Hill Farm in 1949.

Rick separates cow-calf pairs in the corrals
near the old barn at Centennial Hill.

Kyle checks on an electric fence that shorted out
when snow laid prairie grass across the wire.

Rick rests briefly while herding cows and calves on property
near the new house he built in the early 2000s.

Kyle holds a calf's tail out of the way as Jesse applies the
R Diamond brand and the veterinarian prepares the inoculations.

Dave starts the motor on the soybean seed tender, while Rick directs the hose into the hopper at the back of the planter.

Kyle sprays a mix of herbicides and pesticides onto a field of newly emerged soybeans.

The search for a lost dog turns into a social gathering in
the middle of a country road south of Central City.

Every summer on the anniversary of his death,
Meghan returns to the home of her high school boyfriend
who was killed in Iraq in 2006.

Rick makes out a check to the sale barn after entering
winning bids on a pair of black Angus bulls.

In the late summer, when he comes to load up cattle in Curtis,
Rick always visits the farmsteads where he grew up.

near a fence line or had strayed off property. But other incidents were clearly more planned out. Some law enforcement
officials said that the crime wave was fueled by the meth
epidemic—a strange confluence of rising rural addiction and
soaring sale barn values for cattle. When a calf can go for
over a thousand dollars, Rick explained, the temptation is too
great and the risk of loss too severe to let yourself feel squea-
mish about the prospect of branding. There's a reason for the
old saying: "Trust everyone, but brand your damn cattle."

So one by one, Rick used the herding paddles to guide the
young males down the chute. As each calf was clamped into
the cradle and swiveled onto its side, the vet swiftly admin-
istered a series of inoculations, and Jesse pressed the electric
brand, a large R Diamond, into its right flank. Jesse cut an
unlikely figure for the job. His long curly hair was bunched
into a ponytail and tucked through the hole at the back of his
camouflage Pioneer baseball cap. He wore a bright red Ver-
mont t-shirt that read JEEZUM CROW and white wraparound
ski sunglasses to shield his eyes. Each time he applied the
brand, thick billows of white smoke instantly engulfed him,
and the air filled with the smell of burning hair and cauter-
ized flesh. The calves usually let out a single, low moo then
dropped silent. Right after, Kyle would slice into the calf's
scrotum, cut out its testicles, and toss them aside into the dirt.

Later, after the task was complete, I asked Kyle if the branding and castrating bothered him at all or if he just regarded
it as a necessary task. "I really don't think the branding hurts
them too much," Kyle said. "I mean, it burns, but then it cau-
terizes the wound immediately." He said that they do freeze
brands with identification numbers on the flanks of the cows
when they get bigger, using a brand dipped in liquid nitrogen.

The cold cauterization creates a white brand that stands out against the black hide of the Angus cattle, making it possible to pick out specific cows even at night or if they lose their ear tags. "They complain about that more than the hot brand," Kyle said. "And the castration—well, I can't imagine that it feels too good, but I just try to be quick about it. Some people use a bander that applies a tight plastic band at the base of the scrotum, so that everything just drops off after a few days of the blood not circulating. I don't know if that's more humane—some people think so. For us, we think it's better to just cut them, so it's quick and clean, and if we get them out onto grass right afterward, there's pretty much no chance of infection."

After another hour of branding and cutting, Rick stretched his arms over the rails of the corral, letting his gloved hands dangle loose as he caught his breath from chasing calves into the chute. "You know, everybody worries about the branding," he said, "but if I had to choose between being branded and being castrated . . ." He ducked his head, so he had me clearly in his sights. "Well, which would *you* choose?" As he chuckled to himself, Jesse called out the new calf's tag number. Meghan marked it down and then stomped off without a word. She pulled off her yellow stocking cap and crammed it in her back pocket, dangling like a penalty flag.

Later, she explained that Rick had gone to the trouble to move away from hormones and antibiotics used by other ranchers to bulk up their animals, putting the family in a position to charge a premium for their beef, but hot-branding is strictly prohibited if you want to get the top dollar that goes along with Animal Welfare Approved certification. Because many small producers in Nebraska were applying for that label—as well as organic, non-GMO, kosher, and other specialty labels—the Hammonds struggled to sell their beef directly to consumers looking to buy from independent farmers. So most of their cattle, despite being grass-fed, hormone-free, and antibiotic-free, wound up at the sale barn with everyone else's cattle, often to be purchased by Cargill or Tyson or some feedlot that fattens its cattle on corn and growth promoters to sell to those packers. At auction, Rick could command a small premium because his cattle were designated as NI—no implants—but that higher price would be offered only because the buyer knew those cattle would respond to hormones and bulk up faster when they were injected in the feedlot. To Meghan's mind, it was a lot of extra work for the Hammonds, keeping their herd healthy and drug-free, only to have their cattle wind up just like all the other cattle out there.

Rick saw hot-branding in just the opposite light. Not only does it protect your herd from rustling, but when you go to the sale barn, the brand on your cattle becomes your brand as a company. In the western part of Nebraska, brands, checked and approved by government inspectors, are required to sell all cattle; in the eastern part of the state, they're not required or checked, so it's not at all unusual to see unbranded cattle. Rick told me that his cattle, with their distinctive brands burned into their rumps, sent a message to sale barn buyers that his cows had been raised in a traditional way, spending half their lives on open rangelands in western Nebraska. And he didn't deny that the practice made him feel connected to a time when cattle grazed over broad expanses on the unfenced prairie and had to be branded with irons reddened in the coals of a campfire just to assure that professional marauders didn't make off with a rancher's livelihood. "It's the cowboy mystique," Rick said unapologetically, and he said that he considered that way of raising cattle part of the R Diamond identity.

As each brand was set, Jesse rotated the cradle back to vertical and released the trap with a foot-operated pedal, letting the calf leap out into a holding pen with the others. When all the new calves had finally been inoculated, branded, and castrated, we all sat on overturned buckets in the narrow shade of the barn, eating sandwiches and drinking sodas. Meghan explained that they expected to have maybe 130 calves. Fifty would go out to pasture for the summer near Valentine, in the northwest corner of the state, and the other 80, split up into two separate trailer loads, would go out to Rick's land around Curtis. The calves they had branded and castrated that day were ready for pasture now. Soon they would be loaded up for western Nebraska.

"And they're all new calves—from this year?" I asked.

"Yeah, they're only two months old by the time they go," Meghan said.

I studied her a moment, trying to gauge her patience. She had finally ditched her stocking cap for good as the day grew warmer, and the bustle and unease of branding seemed to be draining out of her.

"You know," I ventured, "from my perspective, the branding and all, it seemed like it went—"

"It was good," Meghan interjected. She leaned over and slugged me in the shoulder, as if to signal that the subject was closed. "It's hard on the calves, and it's hard on the people," she said. "But, yeah, otherwise, it was good."

PART THREE

SEEDS OF CHANGE

May 2015

D ave flipped open the lid of a yellow hopper at the rear of the John Deere planter, then lifted a 20-kilo bag of seeds and balanced it in the crook of his elbow. The bag was made of brown paper, like a grocery sack, and printed with a pale green stripe and bold lettering: FEMALE PARENT CORN. Stitched to the top were two coded identification tags listing the batch number, the lot number, where these parent seeds had been grown, and the exact number of seeds per kilogram. Overhead, the sky was slate-gray and threatening, but the steady rains of the last week had finally relented enough to get the seed corn planting underway.

Dave pulled a tab at the bottom of the bag, unzipping the seam like pulling a ripcord, and the seeds poured out until the hopper was full. The turquoise-tinted kernels inside had been treated with a mix of insecticides and fungicides developed by DuPont Pioneer chemists to kill off beetles and aphids and to knock down infections that can slow seed emergence and leafing out. The coating's candy color is just a dye, a reminder to farmers that their seeds are covered with potent chemicals and should be handled with care. But Dave wore a plaid Western shirt, torn off at the shoulders, no sleeves, no gloves, nothing more than his glasses and a Pioneer cap for protective gear.

Once the individual hoppers were filled, one for each planting row, Dave emptied the last of the bags into the bulk bin, then climbed into the tractor's cab. He started the onboard computer, and a fan whirred to life, pressurizing the planting system, sucking up kernels, and sending them rattling through a series of tentacle-like hoses, each extending to one of the small square bins. Underneath, a V-shaped pair of opener blades was set to cut a trench to precisely the right depth, followed by a nozzle from the mini-hopper to drop seeds according to an infrared sensor, and finally a close wheel that would cover the seed and pack the soil to the appropriate density. The hoppers stretched for a dozen furrows in either direction, but on that day, Dave had the system set for planting seed corn, so for every four rows of female seed, there would be a blank row to fill in later with males.

Seeing that everything looked good, Dave eased the planter to the edge of the field, lowered the blades, and started, slowly, across the acres of dark soil, watching over his shoulder to make sure the right pattern was unfurling behind him. Ordinarily, the Hammonds do no-till planting. With no-till, the farmer plants right through the stubble of last year's crop and the tangle of subsequent weeds, using a coulter blade, sharp and wide enough to slice through roots and open a seed trench. By leaving those root systems in place, farmers can reduce erosion and topsoil loss—and, ideally, reduce the use of chemical fertilizers. But for some reason, Dave wasn't really sure why, Pioneer still preferred tillage to prepare a bed for its seed corn fields. So Rick had gone over the field with a one-pass soil finisher, a massive contraption that unfolds into a twenty-four-row, pull-behind tiller that combines a line of discs and a set of miniature coulters. "You can see it's been worked," Dave said, pointing out the distinctive corduroy lines. Behind us a plume of dust swept east across the field, rising like a growing whirlwind, even after all the spring rains.

As we moved, Dave showed me where the planted seeds

were filling in the virtual field on the touchscreen and how the planter was varying the density of seed populations according to the pre-programmed prescription for the field. Apex, the software created by John Deere to control everything from the GPS-guided autosteer to planter row shutoffs, made it so that he didn't have to do anything but get the planter to the appropriate starting point and hit go. The Hammonds began using the system about five years before, but he still hadn't warmed up to the change. "I would rather do it here," Dave said, "but a lot of it, it won't let you. You got to go through Apex, and I can't figure out the goddamned thing. So Kyle, he helps me get it all set."

Months earlier, when the plans were still being drawn up, Kyle had shown me how this field was mapped and entered into Apex. It was cold outside that day, the space heater in the shop at Centennial Hill buzzing with effort, but Kyle was hunkered down at the computer, double-checking everything. "Rick told me to sit down and figure out what we were going to plant for seeds, all the herbicides for the fields, the fertilizers," he said. "That's pretty big, you know? Especially with prepaying. It's kind of a tough thing to do cost-wise, because we're still in the middle of selling. I'm trying to do research and get the best plan put together. It's definitely more responsibility. It's, like, a $300,000 purchase that he pretty much isn't overseeing at all."

To an outsider, it may look simple to raise row crops: just plant a seed and watch it grow. But with commodities prices still under the cost of production—corn stalled under $4 per bushel, soybeans under $10—farmers were making tough decisions about where to spend on inputs. A survey by the *Progressive Farmer* in Omaha found that landowners were planning to invest less in hybrid seeds and scale back as much as possible on fertilizer and chemicals for weeds and pests.

"It's all a trade-off," Kyle said. "Soybeans are still doing better than corn, but if you're not rotating, then you're losing fertility. So you make more this year, but maybe you're shooting yourself in the foot for next year." The stakes were especially ratcheted up by the University of Nebraska's annual survey of farm values, which showed that properties across the state had taken a hit. Even irrigated acres in the north-central region, which contained York and Hamilton counties, were off. So the collateral backing everyone's loans was losing value—raising the specter of what could happen if debts were ever called in as they had been a generation ago. It was a high-risk moment for Rick to hand over the reins. Kyle understood what a vote of confidence it was, but he also knew that he couldn't afford to screw it up. And right now, the top money-maker was the seed corn fields. He had to do them right.

"This light gray area is the isolation zone," Kyle said, showing me an empty spot on the western edge of the field. "So what they're saying is: you have to be 165 feet away from the neighbor's yellow corn. That's actually a Nebraska state law, to have at least that much setback so you don't get cross-pollination. And you can put whatever you want in that isolation, as long as it's not corn. So we plant soybeans there. And then there are fourteen rows of male corn. That's what's called border rows. They help ensure pollination." Finally, down the center of the field map was an alternating patchwork of four rows marked for females and single rows for males.

This whole process of forced cross-pollination in order to produce robust seed corn is arguably the most important development in the last century of agriculture. For all the talk about tractors and center-pivot irrigation systems, the emergence of reliable seed, more than anything else, has turned cycles of catastrophic loss and foreclosure into a reliable if ever-changing business. A hundred years ago, my great-grandfather L. C. Genoways was one of the most successful seed corn producers in Hamilton and York counties. L.C.'s seed corn received "most of the important prizes" awarded by the Nebraska Farmers' Institute in its early years and by 1930 was reported to be "sought after far and wide," but corn seeds in those days were all coming from open-pollinated populations, which means that even seeds that came from the same ear were each genetically distinct. So much genetic variability made it impossible to steadily improve yields, because selecting the best seed came down to eyeballing kernels and hand-selecting based on experience of past years. Even sophisticated breeders, like L.C., who knew how to cross-pollinate, would breed two varieties one year and get large, healthy ears, only to find that the same two varieties produced scraggly cobs with missing kernels and dead tips the next. So if you look at USDA records for corn yields from the end of the Civil War to the beginning of the 1930s, you'll see that—despite huge advances in planting and harvesting equipment, fertilizers, and pest control—the per-acre average remained right around twenty-five bushels for more than sixty-five years.

That all changed with the arrival of a corn breeder who one plant geneticist at the University of Nebraska told me was "the Bill Gates of the seed industry."

HENRY A. WALLACE, known to friends and family as H.A., was the son of Henry C. Wallace, the longtime president of the Corn Belt Meat Producers Association and a liberal-minded professor at Iowa State Agricultural College who counted George Washington Carver among his students. Carver took young H.A. on daily walks, teaching him about parts of flowers and the reproductive processes of native grasses. As Carver gained international fame at the Tuskegee Institute for finding new uses for peanuts as a way of supporting poor, small-plot African American farmers in the South, Wallace was beginning his own studies at Iowa State and dreaming of similarly transforming corn production for poor white farmers in the Midwest.

When he was still a student, Wallace began experimenting with hybridization. Rediscovering Gregor Mendel's groundbreaking research on pea pods, he hit upon the key insight that the only solution to overcoming corn's genetic variability was to first produce consistent varieties, with each kernel the same, that could be used as "parents," year after year. It was possible to achieve this level of artificial selection because of two peculiarities of corn plant anatomy and reproduction. First, corn plants are all naturally hermaphroditic. The male flower of a corn plant is called the tassel; the ear with its corn silk is the female flower. Under normal conditions, the silks of a corn plant are pollinated by the tassels of neighboring plants, but if pollen from other plants isn't available, corn has a second peculiarity: the hermaphroditic parts are capable of self-pollinating.

Of course, when a plant self-pollinates, the progeny very closely resembles the parent, because you don't have DNA introduced from other individuals. In nature that kind of inbreeding would be detrimental, because it could lead to reinforcement of undesirable characteristics (like the Habsburg jaw or Romanov hemophilia among the intermarried royal families of Europe). However, if you are selecting for a desirable characteristic and forcing progeny to self-pollinate for multiple generations, you eventually wind up with a plant that is so genetically pure that the desired trait can be reliably passed from one generation to the next.

Wallace worried, however, that planting inbred varieties was a dead end, because undesirable traits were often transmitted along with the desirable ones—so you might end up with a cob with consistently dense kernels, for example, but the ear itself turned out to be underdeveloped or the plant didn't grow to full height. But then, in 1920, Wallace found a paper by Donald Jones, a chemist at the Connecticut Agricultural Station's experimental farm, describing how he had successfully inbred two separate varieties of corn and then crossed them to produce a durable, high-performing hybrid. Wallace recognized that this was the key to creating seed corn with consistently higher yields. You could build a stockpile of inbred seeds with desirable characteristics and then, year in and year out, plant one inbred variety side by side with the other. Just before flowering, the tassels of one would be removed to make a silk-only "female parent" to receive the pollen from the intact tassels of the "male parent" rows. This crossbreeding produced outstanding kernels to use as seeds for the following year.

This is still essentially the crossbreeding technique employed by all major seed producers, but it's not quite so simple as planting two varieties in the same field. To achieve proper pollination, farmers have to calculate the "nick"— the period when each plant will be sexually mature. But because they are separate varieties, the plants reach maturity at different rates, requiring different planting dates for the silk parents and the pollen parents to ensure that the silk emerges at the same time that the pollen is shed. To hit that narrow window, seed corn farmers use a final piece of knowledge about corn.

The plants mature according to the approximate amount of heat they receive, starting from planting. "It all goes by growing degree units, GDUs," Dave told me. One GDU is added for every degree the average daily temperature exceeds 50°F. So, for example, a day with an average temperature of 65° would be counted as 15 GDUs. "And they know how many GDUs it takes for the female hybrid to get to pollination and how many GDUs it takes for the male to get to pollination, so that nick happens at the same time," Dave said. "If you put them all in at once, your male could shed pollen before the female has started to silk, and you get no pollination. If that happens—and it has—you can lose a whole crop."

"That's why weather is such a big deal," Dave continued. "You plant your males and females days, maybe weeks, apart—all depending on the temperatures. You get a string of days like we just had, where it never gets above fifty degrees, and you just have to wait." On days like those, there's an even greater complication. Several times Rick had described to me the bizarre growing season of 1991—when, halfway around the world, Mount Pinatubo erupted in the Philippines and, in his words, "put a whole shitload of ash in the upper atmosphere." In Nebraska, temperatures climbed like normal in July and August, but the solar radiation that causes corn to grow was diffused. "That was the first year that I'd ever read of a measurement called Langleys," Rick said. Instead of just measuring heat, Langleys measure the density of energy distribution, capturing the difference between a muggy overcast day and a cloudless day when the sun is beating down. "It was cloudy that year, just like this year's been," he said, "and everybody got into their fields at harvest, and it was twenty bushels off what they were expecting."

"When it gets overcast," Dave explained to me, "it can be a crap shoot on getting the weather right." As he reset the planter and got us headed south again across the field, he said that there were practical concerns beyond the problem of estimating the nick. "This field over here," he said, pointing to the neighboring quarter section. "They wanted us to plant last week, but we kept looking and, God, they were calling for rain all week. What if it heated up and the corn matured, but it was too muddy to get in to plant the other rows? So we had to wait—especially with that field, because that was male put in first. You know, my male row planter I can pull around in the mud a lot easier than this thing." He pointed a thumb back at the rows of hoppers, all heavy with seed and bouncing as the planting blades sliced into the loose soil behind us. "If it's wet, I can't do anything with this."

When Henry Wallace had seen how complicated creating reliable seed corn would be, he started to envision a nonprofit organization, run with government cooperation and potentially even public funding, that would distribute his seeds to midwestern farmers for free. He was a man of unusual commitment to the common good and wrote a friend that he did not consider himself a corn breeder but rather "a searcher for methods of bringing the 'inner light' to outward manifestation." In 1921, Wallace's father was appointed secretary of agriculture by President Warren G. Harding and might have spearheaded such an effort. But barely two years later, his father died unexpectedly at age fifty-eight, and Calvin

Coolidge, Harding's successor in the White House, soon settled into the laissez-faire years of our nation's history. Seeing little chance of an ambitious national program gaining traction, Wallace started the Hi-Bred Corn Company, the world's first hybrid seed producer, as a private business in May 1926. But the path to fortune was anything but swift.

Initially, the company saw little in the way of profits. After all, Wallace was asking his seed representatives to try to convince farmers that his seeds were superior to any of those they had been using for years. So Wallace's lead salesman, Roswell Garst, came up with a daring plan. He went from one farm to the next, across sixteen counties in western Iowa, giving away enough eight-pound sample bags of Hi-Bred seeds for farmers to plant half their fields. Whatever additional yield the hybrid corn produced, he would split fifty-fifty with the farmer. After several years, farmers realized that they would see greater profits by simply buying the bags of seeds, rather than sharing the surplus yield with Garst.

Those shared harvests generated income for the young company, but they also produced something more important: they returned a fount of information about how Wallace's seeds performed under different growing conditions. Now renamed Pioneer Hi-Bred, the company turned a sizable chunk of its revenue back into research, hiring a team of new corn breeders to devise still more hybrids that would show improved performance under tough growing conditions, such as the shortened seasons of cold-weather climates or in dryland areas where farms were without irrigation. In the early 1930s, Perry Collins, one of Wallace's researchers, developed Hybrid 307—the first corn specifically designed and marketed for drought resistance, hitting feed stores just as the country spiraled into the Dust Bowl. And when Wallace was, like his father, appointed secretary of agriculture by Franklin Roosevelt in 1933, he finally had the resources to nationally evangelize for hybrid seed, which he believed had the potential to rescue the nation from the Great Depression.

The transformation that followed was staggering. When Wallace joined Roosevelt's cabinet, less than 1 percent of America's corn came from hybrid seeds. A decade later, more than three-quarters of all corn was grown from hybrids—and Pioneer seeds steadily produced the most impressive results. To keep record yields from depressing corn prices, Wallace created the "ever-normal granary," under which the government would establish a federal grain reserve. In years of high production, the Department of Agriculture would buy corn and store it to keep prices up. In years of crop loss, the government would release the reserve to keep prices down. Wallace's plan was hugely popular, stabilizing American food prices— and winning him a spot as FDR's running mate in 1940.

As vice president, Wallace convinced Roosevelt to go even further. To prevent runaway production and the overplanting that created many of the Dust Bowl conditions, Roosevelt signed the Soil Conservation Act, establishing subsidies for farmers to restore native grasses and trees, rather than planting commercial row crops that reduced groundcover and depleted soil nutrients. At virtually the same time, FDR authorized employing federal foresters to plant more than 200 million trees at the perimeters of farm fields in a hundred-mile-wide zone from the Canadian border to the Brazos River. The idea was to create a national windbreak— what was termed the Great Plains Shelterbelt—to reduce the velocity of dust storms and the loss of moisture due to evaporation from windswept soil. And, in fact, soil loss had been reduced by 80 percent by the end of World War II, even as crop production soared.

But Wallace's remarkable Hi-Bred Corn had one significant drawback: it consumed far more nitrogen compounds from the soil than ordinary corn—more, in fact, than almost any other crop. During the war years, the government solved the problem of depleted soil by simply putting more acres into production, but after World War II, the Department of Agriculture found a different solution. Giant chemical manufacturers, like DuPont and Monsanto, had secured wartime defense contracts to produce ammonium nitrate and anhydrous ammonia to make bombs and other munitions. Now, they argued to the USDA, those chemicals could be used as fertilizers.

✺

The Miller Nitro sprayer, a cherry-red colossus on 6-foot-tall all-terrain tires, rolled up to the edge of the freshly planted field of corn. The twin booms of the spray rig were spread out in either direction, like a pair of enormous dragonfly wings, unfolded and lowered by a system of hydraulics. The corn was just tall enough that the first sprigs of green were showing, and Kyle was preparing to spray them with a soup of liquid herbicides to wither any weeds that might compete for nutrients and sunlight. But before Kyle turned on the sprayer, he wanted to reassure me. "The cabin is completely sealed," he said. A pressurized ventilation system with built-in charcoal filtration and a secondary in-cab filter would make sure that none of the chemicals in the sloshing 1,000-gallon tank behind our seats, which Kyle was about to turn into a low-pressure mist, was going to make it inside. "I'm not going to fumigate you," he said with a laugh. "I promise." With that, he flipped on the sprayer and started bouncing across the field, much faster than I'd imagined.

Before getting to this stage, Kyle had made a pass just before planting with the same machine, applying liquid nitrogen for the N-hungry corn hybrids. When Henry A. Wallace created the first inbred corn varieties, he hadn't anticipated one problem: uniformly healthy corn plants that experienced rapid growth—going from knee-high to shoulder height in as little as two weeks—swiftly and voraciously converted nitrogen from the soil to proteins needed for green tissue growth. As new hybrids yielded taller stalks, now growing as much as 10 feet tall, and were bred to be tolerant of tight planting, allowing farmers to plant more seed per acre, the natural nitrogen content of the soil proved too little to sustain a crop through a full growing cycle. Thus, the application of nitrogen compounds has become central to modern, industrial-scale farming, but the discovery of its importance was the result of a bizarre happenstance.

As early as 1910, the DuPont Powder Company, a munitions manufacturer so large that they are estimated to have provided half of all the gunpowder used by the Union Army during the Civil War, had sought to expand its business by encouraging the use of its explosives on midwestern farms. Company representatives distributed a free pamphlet, entitled *Farming with Dynamite*, which promised to simplify the tasks of removing stumps, clearing land, and breaking up hardpan. Two years later, Edward Lewis, a DuPont representative addressing the Nebraska State Horticultural Society, acknowledged that "the name of DuPont suggests something rather scary," but he said that a farmer in Thayer County, Missouri, who had dynamited his entire field had made a discovery exciting enough to ease their fears. "He planted his corn in the land that he had dynamited," Lewis reported, "and it doubled his crop the first year."

Of course, it wasn't the breaking up of the soil that was promoting rapid growth but the nitrogen compound contained in gunpowder. In 1913, Fritz Haber, working at the Oppau plant of the German chemical company BASF, succeeded in developing a less destructive way of achieving the same result: "fixing" nitrogen from the air by combining it with hydrogen into ammonia and then catalyzing it. The resulting nitric acid remains the basic building block for the manufacture of all modern fertilizers—but Haber's process also ushered in a new era of chemical weaponry. During World War I, Haber not only devised more powerful explosives but also poison gases, such as ammonia, then chlorine, and eventually the Zyklon B gas used on Jewish prisoners in Nazi concentration camps.

By that time, the use of chemicals against humans had been outlawed under the Geneva Protocol, so American companies like DuPont found other ways of refocusing their laboratory know-how on the war effort. The herbicide 2,4-D was developed as a potential destroyer of enemy crops. The insecticide DDT was manufactured to prevent the spread of typhus-carrying lice among GIs. But as soon as the war was over, DuPont turned to marketing those same chemicals as DuPont Garden Dust and DuPont 5% DDT Insect Spray. Company advertisements touted these products as "Better Things for Better Living . . . *Through Chemistry.*" But gardens were just the tip of the iceberg. DuPont, along with other giant chemical manufacturers like Dow and Monsanto, teamed up with the old grain cartels, including Cargill and ADM, to lobby for congressional support for producing these compounds as large-scale agrichemicals.

In 1953, the industry found its greatest ally, when Ezra Taft Benson took over as President Dwight D. Eisenhower's secretary of agriculture. (Wallace, by then, had retired from public life after making a failed bid for the presidency in 1948.) Benson, a high-ranking member of the Mormon Church and a fanatical Red Scare Republican, immediately informed Eisenhower that he was philosophically opposed to the government price supports developed by Wallace, because, to his mind, they were tantamount to socialism. He publicly referred to small farmers as "irresponsible feeders at the public trough" and vowed to return to a system where the greatest profits went to the largest producers. Small farmers hated him for it, but Benson was unflagging. In 1957, at the National Corn Picking Contest in South Dakota, he delivered a speech on the evils of federal farm aid despite a hail of eggs from angry farmers.

Foreshadowing today's aggressive, pro-corporate agricultural policies, Benson argued that the only way to outcompete the collective farms of the Soviet Union and Red China was to use our superior corn and chemical technology to the fullest. The United States could, if it chose, overproduce corn to drive down international prices, and it could use the surplus as a tool of diplomatic leverage in the form of foreign aid. Instead of guns, the United States began to give our allies grain—transforming, for the first time, a food product into a weapon in the national arsenal.

When Soviet premier Nikita Khrushchev visited the United States at President Eisenhower's invitation in 1959, the

Cold War for food dominance started to heat up. Khrushchev specifically requested to see only one man: Roswell Garst, the former Pioneer seed salesman for Henry A. Wallace, who was then head of Garst & Thomas Hi-Bred Corn Company. Khrushchev had met Garst once before, when he visited the Soviet Union, and had become obsessed by the potential of hybrid corn. Khrushchev and his wife spent a day at Garst's farm near Coon Rapids, Iowa. In his memoirs, Khrushchev later wrote, "Garst gave me an entire lecture on agriculture," in which he earnestly explained that American farmers had stopped worrying about crop rotation. Garst told the Soviet leader, "Science today considers that approach outdated. And I think so, too." In past years, planting the same crop repeatedly would have attracted pests and depleted the soil of nitrogen. "Now there is no such problem. We have herbicides and other such chemical substances that make it possible to combat pests," Garst said. And there was no longer any need to plant clover or alfalfa to accumulate nitrogen. "It is more profitable for me to buy nitrogen, potassium, phosphorus, in mix form, and add this fertilizer."

On that same official visit, Benson led Khrushchev on a tour of the U.S. Agricultural Research Center in Beltsville, Maryland. Benson, in his official remarks, said that there was a "constant give-and-take of information between government scientists and those in private industry," adding that "we are all working together within the framework of our capitalistic free-enterprise society to benefit our farmers, all our citizens, and people throughout the world." He listed hybrid corn first among the achievements of such cooperative efforts and introduced white-coated lab researchers who extolled the virtues of 2,4-D and chemical fertilizers. Khrushchev was unimpressed by the visit he made to a farm owned by President Eisenhower, dismissing it as "not on a scale such as we have at our collective farms and state farms." Benson later remembered that Khrushchev bragged, "We won't have to fight you. We'll so weaken your economy until you fall like overripe fruit into our hands." Benson vowed that American farms would outproduce the Soviets through superior chemistry.

By the end of the Eisenhower era, however, environmentalists began to raise concerns about the hundreds of commercial herbicides and pesticides being applied to American crops in quantities totaling hundreds of millions of pounds. In fact, a citizens' group, led by biologist Robert Cushman Murphy and Mayo Clinic hematologist Malcolm Hargaves, brought suit against the USDA in 1958 to bring a halt to indiscriminate DDT spraying in New York State. Benson admonished doubters that "abandoning the use of chemicals on farms and in the food industry would result in an immediate decline in the quantity and overall quality of our food supply and cause a rapid rise in food prices paid by the consumer." Later, when Rachel Carson, who attended those trial proceedings, documented the dangers of DDT and 2,4-D, including elevated incidence rates of rare forms of cancer, DuPont responded by opening a confidential file on her and intimating to her publisher that they might sue. But Carson's message was already out. Her modest calls for mitigating the use of herbicides and pesticides were instead carried out as an outright ban on DDT, the passage of the Clean Air Act, and eventually the establishment of the Environmental Protection Agency.

Before long, Monsanto returned to manufacturing 2,4-D for its original purpose. During the early stages of the Vietnam War, Secretary of State Dean Rusk argued to President John F. Kennedy that the use of chemical defoliants was not technically a violation of the Geneva ban on chemical weaponry, because the chemicals were meant to be sprayed on dense jungle foliage, not enemy combatants themselves. Willing to be persuaded, Kennedy authorized the covert Operation Ranch Hand. Using chemical herbicides provided by Monsanto and Dow, most of them based on 2,4-D or its descendant 2,4,5-T, the military created a series of Rainbow Herbicides, code-named according to different colors on the spectrum. There was Agent Purple, then Agent Green and Agent Pink. Then in 1965, government chemists developed a potent mixture of the esters of 2,4-D and 2,4,5-T, which they called Agent Orange. The herbicide was unusually effective— especially when it was deployed at fifty times the concentration that either chemical had been used in agriculture.

Rick told me that he thought that allowing 2,4-D and other herbicides to be used in that way had been a mistake. The military had been applying those chemicals in concentrations that scientists at Monsanto and Dow had warned would be lethal or debilitating to humans. In time, the government admitted that Agent Orange had led to lung and prostate cancers for American veterans and poisoned the water and food supply of forest villages in Vietnam, leaving behind a legacy of horrifying birth defects. More than that, the fear caused by seeing the effects of military-grade applications of Agent Orange had turned the public against it as an herbicide in the United States. But Rick believed that limiting its use had only increased the production of other, less-tested replacements.

In 2012, for example, high levels of atrazine, a substitute for 2,4-D widely used by corn growers, were found in the water supplies of thousands of municipalities across the

Midwest. Syngenta, the manufacturer of atrazine, agreed to pay more than $100 million to those utilities for the installation of filtration systems, as part of the settlement of a class-action lawsuit. The company admitted to no wrongdoing, but the Natural Resources Defense Council was quick to note that Syngenta's pay-out came after researchers in France linked drinking atrazine-laced water during pregnancy with increased risk of delivering a low-birth-weight baby—with all of the associated health problems for the infant. And that's just one example among thousands of herbicides.

"Bill Moyers had this deal on chemicals on PBS," Rick told me. "They tested his blood and found eighty-four different chemicals there—of which about fifty or so hadn't been invented in 1940." He noted that there are now more than 75,000 chemicals registered with the EPA, but that the vast majority have never been tested to find out their effects on human health. For Rick, and anyone working on a farm, these concerns are anything but abstract. In the waning days of harvest in 2014, the National Institutes of Health had published the results of a twenty-year study that looked at rates of depression among farmers, showing compelling evidence that use of seven different pesticides, mostly neurodisrupters, had nearly doubled the suicide rates among the men who applied those chemicals. Worse still, the depression rates were even elevated among their spouses and other immediate family members. Soon after, another study found two hotspots for Parkinson's disease in Nebraska, apparently linked to historical use of pesticides in those areas.

"You can't ignore it," Rick said, "but banning old chemicals just puts industry on the hunt for new ones—which will be used before we even know if they're safe or not. We can always see the short-term benefits, but we don't see the long-term ill effects until it's too late." That's why Kyle waits for a windless day to use the sprayer, applies as little of each chemical input as possible, and stays inside the cab until a whole field is sprayed and the air has cleared. It's why Rick showers and washes his clothes right after applying these chemicals and insists that Kyle do the same. It's also why Rick fiercely defends the use of anti-pest varieties of row crops that have been developed using genetic modifications. And it's why he eventually decided to plant crops with engineered resistance to a newer product developed by Monsanto as an alternative to 2,4-D-based herbicides.

That herbicide, which has gained widespread use in the last twenty years, is called glyphosate—or, more commonly, Roundup.

RICK HELD off on planting soybeans through most of May. He had chosen to go with a short-season variety again that year and didn't want to get them in too early, risking another wet fall and another late harvest—especially with what he was seeing in the commodities markets. Rick had spent the winter emptying his bins as slowly as possible, keeping back reserves, waiting. There's typically a spring bump in prices, what's called a "seasonal rally," at the start of planting when stockpiles are at their absolute lows and the results of the South American harvest start to come in. "Eight out of ten years, you'll see it," Rick said. "But then a couple of weeks ago, we had a report come out that made everything go the other way. The U.S. had had a big bean crop, and then South America had a fantastic year." Instead of the usual spring bump, prices had actually fallen further on the news. Rick decided to sell off what he had left—about 5 percent of his overall corn crop and fully 25 percent of his soybeans, grain he had been storing for more than six months. It was time to cut losses and hope for a better year ahead. "Maybe I sold on the low," he said. "But I just didn't see upside anytime soon."

So Rick cleared out the last of his bins, hauled the final semi loads to the elevator, and started picking up new seeds from the dealer. But by the time he had gotten the John Deere planter loaded on the flatbed and brought over from the seed corn fields, a line of spring storms had rolled in, blanketing the sky for days. By then, there was nothing to do but put his seeds in as the weather allowed, one field at a time. On the day Rick was ready to move to the quarter section neighboring Centennial Hill, the clouds were stacking up, billowy and roiling. They kept threatening big rain but never produced more than a light mist, so he planted half of the field with one variety, and when the weather seemed like it would hold for a few more hours, he hitched up the seed tender, a portable tank that looked like a miniature water tower on wheels, and headed to the seed dealer to get a second variety to finish the field.

As we drove, Rick explained that his Pioneer dealer offered thirty-eight varieties of soybeans. Almost all of them were genetically modified by bioengineers, inserting strands of DNA that make the plants able to withstand a range of herbicides, particularly the compound glyphosate. Chemist John E. Franz, working for the St. Louis–based chemical giant Monsanto, first hit upon glyphosate in 1970 as part of a company-wide search for potent chemicals that could be sold as weed-killer. Brought to market four years later under the trade name Roundup, the broad-spectrum herbicide saw unmatched growth in the agricultural industry, where it was used for broadleaf weed control. The only problem with Roundup was that it proved so potent that farmers had to apply it carefully, spraying only early weeds, before their corn or soybeans had begun to sprout. Otherwise, it would kill the entire crop.

From Monsanto's standpoint this presented a serious problem. Simply put, the necessity of limiting Roundup's application also limited its sales. So company researchers started to seek out ways of engineering crops that could withstand this new weed-killer. By then, scientists had already developed the revolutionary new ability to cut and splice protein strands into the DNA sequences of bacteria. If they could do the same with plant cells, then, at least in theory, they could chemically insert resistance to insects or to herbicides.

Monsanto had recently entered into a deal with biotech pioneer Genentech to license some of their gene-splicing technology, and they already had successfully used these techniques to produce recombinant bovine somatotropin (rBST), a hormone that extended the lactation period of dairy cows and thus increased their milk production. The hormone, eventually marketed as Posilac, was showing promising results in field tests—and Monsanto executives, now convinced of the commercial viability of genetically modified organisms (GMOs), wanted to find similar modifications that could be made in plants. So company researchers were tasked with trying to find a gene that could be spliced into soybean and corn hybrids to make them resistant to Roundup.

Researchers began by testing every lab sample amassed from the company's fields and facilities nationwide, no matter the reason for its original collection. From a Roundup manufacturing plant near Luling, Louisiana, they received a sample that had been collected from a waste dump, where workers had noticed a range of bacteria thriving despite prolonged exposure to glyphosate. When the sample was taken, Monsanto researchers were hoping to produce a glyphosate-resistant bacteria that could help break down the chemical when there were spills or overapplication events. Now they were looking for the resistant gene inside the bacteria in hopes that it could be spliced into row crops to *increase* the application of Roundup. By 1986, researchers had isolated the particular gene that controlled that immunity, spliced it into soybeans, and were ready to begin field trials.

In the meantime, other Monsanto researchers were trying to find a similar genetic-engineering solution to the European corn borer, an insect that inflicted more than $1 billion in losses in the United States and Canada each year. Since the 1960s, endotoxins produced by *Bacillus thuringiensis* (Bt), a common bacteria found in the soil, had been sold as a commercial microbial insecticide to kill moth larvae. If the specific DNA that produced Bt toxins could be isolated and spliced into corn genetic sequences, scientists believed they could create an ear of corn that would be lethal to the European corn borer but perfectly safe for humans or livestock to eat. Soon, that hurdle had been cleared, and Monsanto began looking for a seed partner to market its pest-resistant corn—and decided to approach Pioneer. If Monsanto could marry its genetic modifications with Pioneer's superior inbred seed stock, Monsanto believed it would have a line of products with unmatched yield potential.

In the 1990s, perhaps too eager to demonstrate the effectiveness of its new GMO crops, Monsanto allowed Pioneer to use its biotech to produce Roundup Ready soybeans and Bt corn—asking only for small usage fees and no royalties. For less than $40 million, Pioneer suddenly had the technology and the sales muscle to move toward genetically modified feed crops. Rather than partner with Monsanto, Pioneer became its greatest competitor, entering into a joint venture with DuPont, which soon came to regard seed technology as so lucrative an opportunity that it bought Pioneer outright (and changed the name to DuPont Pioneer). Despite the rounds of lawsuits that followed, DuPont Pioneer released Roundup Ready soybeans in 1994 and then Bt corn in 1996.

When Pioneer seed reps first started promoting these new products, Rick initially resisted. "I was very reluctant," he said. He objected to Pioneer requiring that farmers sign an agreement promising not to save and clean soybeans to replant the next year. Even though trying to replant progeny seeds from hybrid varieties rarely produces a quality crop, he said that making that mistake should be a farmer's prerogative. "It just felt wrong to me—un-American," he said. "You know, I paid you for this seed. You're telling me what I can and can't do next year with the things that I grow? That's rotten. They say that they went to the expense of doing the GMO modifications. I get that, but you know what? There was expense to seed hybridization before, and we never had to sign anything. It seems to me that if you raise it and grow it you should be able to plant it."

Rick also traditionally spread his risk by buying hybrids from different dealers, but as soon as Monsanto and Pioneer got into cutthroat competition, they each started acquiring smaller competitors, just to expand their market share. "There used to be a hundred seed corn companies," he said. "But it's been winnowed down to a few. The bigs have bought out most of the littles." Unable to compete against the seed giants, small companies relied more and more on the open-source genetics provided by public universities, rather than investing in their own research. Rick began to suspect that he was getting identical genetics, no matter what the label on the bag said.

One year, for example, he planted three different varieties of corn—Super Cross, Maize, and Golden Harvest—and they all fell prey to the same weakness in their stalks. Known as "green snap," corn plants with a particular genetic defect will develop elongated nodes when exposed to herbicides, making them prone to blow-down in heavy winds. Seeing all three varieties display the same problem made Rick suspicious. "I could not physically tell the difference," he said, so he just decided to go with the one company that had always been successful for him. "I really liked Pioneer," he said. "That was it for me. I was done. No more Mr. Nice Guy. There was no sense losing volume discounts, when they were all the same. I wasn't saving money, and I wasn't spreading my risk out." And once he had gone over exclusively to Pioneer corn, it just made sense to plant Pioneer soybean hybrids.

Even so, Rick had his reservations. "I could see that, within a relatively short period of time, if everybody switched to Bt corn, then the bugs would become resistant. And if both the corn and soybeans were Roundup Ready, then you'd end up with weeds completely resistant to Roundup." In fact, in the twenty years after Roundup Ready soybeans were first introduced by Monsanto, the amount of glyphosate that the United States collectively applied each year went from less than 10,000 tons to more than 125,000 tons, and, indeed, the combination of monocultures of corn and soybeans and using a single herbicide for both has produced a range of herbicide-resistant superweeds. "If they had just stuck with Roundup Ready beans and rotated those with Bt corn, using old 2,4-D or other families of chemicals for weed control, then Roundup could have been a viable product for many, many years," Rick said. Instead, farmers tend to use more Roundup and more combinations of other herbicides with Roundup.

Unsurprisingly, the agricultural chemicals team at the U.S. Geological Survey Office, part of the U.S. Department of the Interior, has now detected the presence of glyphosate in the air and water supply in rural states. And the effect of those chemicals is the subject of heated debate. In 2015, the cancer-research wing of the World Health Organization released a report finding that glyphosate is probably carcinogenic to humans. Monsanto hotly disputed the findings. The following year, the UN and WHO issued another report clarifying that, at realistic intake levels, glyphosate was "unlikely to pose a carcinogenic risk to humans from exposure through diet." Still, critics maintain that there are no studies of the effects of exposure over many years. Rick acknowledged this fear about the long-term impact, but he also wondered if it was any worse than the chemicals that had come before.

"It's not like we were all organic farmers until Monsanto came along," he said. He vividly remembered the days of applying dyfonate, a highly toxic pesticide used on corn plants. "You had a separate set of boxes on your planter with this pesticide inside," Rick said. "It was supposed to go in furrow, but invariably when you plant, the wind's blowing about forty miles an hour. In those days, there were no vacuum-sealed cabs. I remember that dust caking up along the sweat band of my cap, and when I took my hat off, it took the top layer of skin with it." Bt corn had done away with many such pesticides, and he said that Roundup had also eliminated many herbicides that were problematic "as far as farmer safety goes."

Besides, Rick said, he couldn't afford to ignore what he

saw while driving up and down these same country roads we were on now: his neighbor's fields of soybeans and corn that looked taller, healthier, more productive than his own. "My decision on whether to go GMO or not was a tough one," he said. "I didn't automatically jump in like everyone else did. But the big deciding factor to me was the old question: wherein lies the biggest evil? Because at the other side, I was using old broad-spectrum poisons—dyfonate, another one called Counter—that were killing all the beneficials. With GMOs, at least, we were using material from naturally occurring bacteria and not just killing everything in sight."

So one year, Rick decided to raise a few fields of Roundup Ready soybeans. "It was a nightmare," he said. "I was like, 'Oh, my God, if I forget exactly where the GMO quit, and I spray Roundup in the wrong places, I'll kill my whole crop.'" Plus, even to plant half his fields with Roundup Ready beans, he had to sign an agreement with Pioneer that curbed his rights on the other half. He was under all of the restrictions without any of the rewards. "And ethically, it's like being a little bit pregnant," Rick said. "Either you are a sinner or you're not." So the next year, and ever after, he bought only Roundup Ready soybeans, and eventually, he bought Pioneer's newly introduced Roundup Ready corn. And in no time, DuPont Pioneer and Monsanto introduced new varieties with "stacked traits," combining Roundup resistance with other desirable characteristics.

At North Forty Seed, Rick's dealer, there's a genetically modified variety for almost any soybean need. One modification causes the plant to create a protein that kills insects; another changes the profile of the fatty acids in the beans, making them less susceptible to rot after harvesting. The *Soybeans Seed Guide* breaks down the performance of each product, all sold under codenames like P22T41R2 and 93Y41, into extremely fine categories: whether they are resistant to particular races of soybean cyst nematodes, to the Phytophthora infections that cause root rot, or to white mold that can cause stem rot; how quickly they emerge, how tall they grow, how wide they canopy; what their protein and oil percentages will be at optimum 13 percent moisture rates at harvest.

The precision is stunning, and the knowledge demanded of farmers, most of whom consult with agronomists but ultimately decide what to plant on their own, seems impossible to fathom—but also impossible to escape. A quarter of a century ago, Roundup Ready crops didn't exist; today, they are 90 percent of the soybean market and nearly three-quarters of the corn and rice markets. And on the strength of that success, Monsanto and DuPont Pioneer have grown into global seed giants, now controlling 45 percent of all the seed sold in the world. Short of going completely organic and dropping out of growing commodity grains, how is a farmer supposed to avoid raising corn and soybeans that have been genetically modified to withstand Roundup?

As we approached the dealership, Rick put on his hazards, pulled past the end of the driveway, and then backed the trailer straight onto the concrete pad next to the seed warehouse. As Rick swung his driver's door open, Dennis Stevens was there to greet him.

Dennis wore his Pioneer cap low, and his snowy mustache nearly covered his smile. "Well, if it isn't the local liberal," he said.

Rick adjusted his cowboy hat as he stepped out of the cab. "If it isn't the local son of a bitch," he replied. The two men smiled broadly and shook hands.

Dennis walked us over to the pro boxes, stacked like over-sized milk crates all the way to the rafters, each labeled with its particular product code but also its place of origin, the date it was field-tested, any preapplied treatments used, and the exact number of seeds in the box. Dennis double-checked which product Rick was picking up with this load, and then gave the signal to the forklift operator to pick up a box full of soybeans and hoist it high over Rick's seed tender. The pro boxes have a trap door on the bottom that, when opened, lets the seeds pour out like sand in an hourglass. The fork-lift revved and beeped as the operator moved it precisely into place. Rick climbed the side of the tender and yanked on the sliding door, sending the seeds echoing into the empty tank. When every seed had drained out, Rick banged hard on the door to slide it back into place, but it wouldn't budge.

"Do you need a hammer?" Dennis called.

Rick gave him a quick, dismissive smile, then pushed on the door again, putting his shoulder into it this time. It slid into place with a resounding clang. "That ought to do it," Rick said.

Before long, we were racketing back over the washboard roads, the heavy tender bouncing and tugging against the hitch. Given everything, Rick seemed oddly ebullient—drumming his thumbs on the steering wheel, whistling a little despite a few raindrops blossoming on the windshield. "I usu-ally have about fifty percent of my beans forward-contracted by now, and this year, I have *none*," he said. He had decided to take an even bigger gamble than the year before, banking on prices going up before harvest. It was an undeniable risk, but after all winter worrying, he was happy to finally be getting back to work, doing something instead of just waiting.

At the field edge, Rick pulled alongside the hopper at the back of the planter. Dave fired up the auger, powered by a small diesel engine underneath that started with a pull cord like a lawnmower, and then Rick, climbing a ladder to the hatch door on top of the main hopper, directed the corrugated plastic hose inside, spreading the seeds out layer after layer. Besides their many genetic modifications, the candy-bright seeds, like a barrel of green M&Ms, had been treated with a mix of insecticides and fungicides similar to those used on the seed corn. Rick wore a plaid flannel shirt buttoned at the wrists and thick rubber gloves as protection. All the while, he kept eying the sky. "I'm afraid we're going to get everything loaded up just in time for it to rain," he shouted.

The prospect of rain was concerning, not only because the planter can mire and get stuck in a muddy field. Even a light rain can affect seed depth for each furrow, forcing the farmer to constantly recalibrate and double-check the planter to keep from wasting seeds. Rick often recalls the spring, not long after he'd started working with Heidi, that he planted too deep and nothing sprouted, so he replanted but too shallow and a spring rain washed everything away, so he replanted again, by then needing a perfect season just to make up for the lost investment in seed. Rick always wore a mustache up until then, but that year, the worry turned his facial hair snow white. In the end, he broke even for that year, but he's been clean-shaven and a bit more mindful of his seed depth ever since.

Dave climbed into the cab of the tractor and started the onboard computer. The fan pressurizing the planting system came to life again. A big chemical tank to one side of the planting attachment made it so that Dave could apply additional fungicide, insecticide, or fertilizer right as the seeds came out of the bulk bin. He pulled the planter parallel with the last completed rows, lowered the planting blades, and started again across the field. For the moment at least, they had beaten the rain.

"They say farmers are the world's biggest optimists," Rick said, watching Dave go. "You have to be optimistic in this business, just to keep going—or at least have an awful short memory."

AFTER ENOUGH warm days that the female seed corn was moving toward maturity, Dave returned to the fields to complete planting the missing rows of male corn. The planter crept forward, the blades opening and closing furrows in the blanks between the female rows, where tiny corn plants now sprouted green and tender. Rick leaned against the bed of his truck, watching. Pioneer had sent out a field manager earlier in the day to check on the progress of this quarter section, and Rick was worried about what would happen if the guy came back and found an outsider there. I assured Rick that I had checked in with the company's media liaison, that the last thing I wanted was to cause him any headaches.

He nodded ruefully, visibly unconvinced.

"Pioneer used to be a wholly owned family deal," he said. "Then it got bought out by DuPont. Since then, I don't know. We've only been planting seed corn for them for seven years, but it seems like a lot of corporate edicts come down now that your local people just have to follow. What the solution is, I don't know. I'm sure they've got their reasons. I just wish it were a little less top-down, a little more grower-friendly."

As one example, he explained that DuPont Pioneer now required its farmers to account for the custody and security of every bag of seed corn—right down to requiring a pallet strap for hauling from the company compound, just outside of York. After all Pioneer has done to secure and protect its intellectual property, the last thing the corporate bosses wanted was to risk a spilled load—seed scattered along the roadway or into the ditch, raising the possibility of some farmer scooping up their patented genetics, or seeds simply taking root and cross-pollinating with other corn plants. And even after you've picked up the seeds and secured them, you have to agree to drive nonstop from the pickup point to your farm. By then, you've already had to prove that you have a locked shop. They come out and check it, and they insist on more than just a barn with a padlock; they want a framed building with a secure door—and the seed has to go straight from their loading dock to that shop or else directly to the field.

As Dave's tractor neared the pad of the pivot in the center of the field, Rick took out his cell phone. "You call them as soon as you start planting," he said, dialing. He told the Pioneer seed representative that Dave was underway on a new field and gave them the identification numbers for the bags and location. That way they could come to check that everything was being planted according to their agronomists' guidelines and then add it to the list of fields under surveillance by the seed security teams.

It sounds like corporate paranoia to have company trucks patrolling the back roads looking for anything suspicious, but in early May 2011, a farmer in the midst of corn planting outside Tama, Iowa, watched as a car pulled over at the edge of his seed corn field. Two Chinese men stepped out and came scrambling up the ditch. One said that they had been attending an international agricultural conference in Ames and wanted to see someone planting a real Iowa cornfield. The farmer was dubious. Ames was nearly an hour away with nothing but expanses of corn in between, all at the peak of the planting season. How had these two men chanced upon his field—on the very day he happened to be planting an experimental and top-secret seed under development by DuPont Pioneer?

It was unusual enough that the farmer contacted his seed representative, who sent out a security team to keep an eye on the newly planted rows. The next day, the field manager spotted the same car. He watched as the passenger got out, climbed the embankment, and then knelt down in the dirt and began digging corn seeds out of the ground. When confronted, the man grew flustered and red-faced. He now claimed to be a researcher from the University of Iowa on his way to a conference. But before the Pioneer field manager could question him further, the man fled back down the embankment. He jumped into the waiting car and took off, swerving through the grassy ditch before fishtailing onto the gravel road and speeding away.

A few weeks later, during a routine meeting with agents from the Iowa office of the FBI, a Pioneer executive mentioned the incident. He explained that Pioneer enters into exclusive contracts with farmers to grow proprietary seeds. The exact genetic sequence of successful seeds is a tightly held secret, worth tens of millions of dollars. The Pioneer field manager had written down the license plate number and handed it over to company security. They had been able to trace the plates back to a rental car company at the Kansas City airport. Representatives there said the car had been rented by Mo Hailong, a Chinese national working as the Director of International Business at the Beijing Dabeinong Technology Group Company (DBN), a company with a seed division in direct competition with Pioneer in China.

The FBI launched a full investigation. Over a period of nearly two years, crisscrossing the United States, they tracked Mo and a team of five corn industry spies, including insiders at Pioneer research labs, from state-sponsored dinners in honor of the Chinese government in Iowa to secret research facilities in Florida to a covert test plot that Mo had established in rural Illinois for cross-breeding proprietary seeds in preparation for smuggling them to China. It seems like a lot of cloak-and-dagger for a few seeds, but biotechnology is now worth billions of dollars per year in sales—and the U.S. government believes that something much larger is going on.

This theft, they argue, stemmed from an undeniable and dangerous fact: despite its remarkable landmass, China simply can't grow enough food to feed itself, particularly now that the country's burgeoning middle class has acquired an appetite for meat. (Most corn in China is used as feed for livestock.) Water shortages and lack of arable terrain have forced their government to buy between 2 million and 5 million metric tons of American corn annually, approximately 94 percent of all corn imported into China each year. If China hopes to feed (and pacify) its growing population while also loosening the very real stranglehold that America has on its national food supply, its farmers have to start producing a lot more corn—not just enough to meet their domestic demand in good years but enough to maintain a stockpile to offset their global market impact during bad ones. The only tenable way for China to meet its own demand is by planting high-performance hybrids, which can single-handedly double or potentially even triple per-acre corn production.

In recent years, DuPont Pioneer has increased its share of the seed corn market in China from less than a tenth of a percent to 12 percent. The company has told Chinese officials that they should Americanize their agriculture: consolidate land, plant GMO seed, apply industrial fertilizers, subsidize the sale of planting and harvest equipment. DuPont Pioneer argues that they are striving only to feed a burgeoning global population. William S. Niebur, the head of the company's operations in China, asked, "Without China's food security, how can we ever imagine an effective, realistic, sustainable global food-security system?" But many, including the Chinese government, expressed concerns that DuPont Pioneer's goal was not global food security or feeding the Chinese people, but rather increasing market share and profit by keeping China as a customer. Whatever the reason, the FBI pursued Mo, who was eventually charged with conspiracy to steal trade secrets from Pioneer and Monsanto and sentenced to three years in prison.

Ever since that case, Rick said that security had grown even tighter. "See the tag on every bag of seed?" he said, pointing to a stack of sealed bags waiting to be opened and loaded into the planter. He said that they used to put a rubber band around those tags and put them in a plastic tube at the field's edge for the agronomist to pick up periodically. "Now those tags go into a manila envelope," Rick said, "and as soon as you get done filling the planter, you call them to get it. A lot of times they'll come out while you're still planting. They check the seed depth and the seed population, but really they're just double-checking to see how everything looks at your field. They've got it figured down to a gnat's ass on how many seeds you should have, and they want to make sure none go missing."

Pioneer is so worried about leaking information and stolen biotech that Rick doesn't even know what variety of seeds he's planting, just its codename. All of his communication with Pioneer refers to that alias and the field number. He said that in recent years Pioneer wouldn't even release another pallet of seeds for planting until the company could confirm that the planter was almost empty. In May, when planting gets going full steam, many farmers are planting late into the night. Rick said that waiting around for seed can add hours to the process, but Pioneer will do everything it can to keep farmers working. "A while back, Dave called at ten-thirty at night, and they ran him out a couple extra bags," Rick said, and when he was done planting at 2 A.M., he called again, and the seed rep said that a field manager would be out to meet Dave first thing in the morning. "As soon as you get your planter cleaned out, they're there, and even if you have a fourth of a bag of excess seed, they want it back."

At the far end of the field, Dave wheeled the tractor around and lined up the planter again, running another set of male rows coming back in our direction. For all the hassle, Rick wasn't complaining about raising seed corn for Pioneer. He was the first to say: it had become the anchor for his whole operation and the one sure source of income, buffeting against bad times, like the current market downturn, when other farmers were forced to rely on commercial crops. "The seed corn is much more lucrative," Rick said. "And it's a lot less work. I mean, we have to plant it, of course. And midsummer, a week or two after they're sure the pollination is completely done, we take the sprayer and attach this big shredder. We go through, and we destroy every male row. That's a little time but hardly any expense. But that's it. We don't do the harvesting ourselves. They have contracted different farmers, three or four guys, to do the harvesting. We have to be there to move the pivots for them. Then there are trucks that are contracted—maybe twenty trucks—to haul it away, and that's it. So we don't pay for any of the seed. We don't pay for any of the harvest. We don't have any storage. And the price is set just under the Board of Trade for commercial corn in whatever month you choose to market it. So it's almost all of the profit with none of the usual expense or risk. It's so much easier."

I asked Rick how they paid out for the seed corn fields. With the inbred varieties producing such small and often scraggly parent plants, you wouldn't think that the yield would be enough to make it worthwhile. Rick laughed hard. "The first year we planted seed corn, we got so worried. The plants were only about yea big," he said, gesturing to mid-thigh. "I went out, and I saw ears half-filled. Oh, man, I thought we were going to be in big trouble. No one told us! It was like, 'Jeez, what did I do?' It turned out we were their biggest producer for that variety that year. They were thrilled."

He explained that DuPont Pioneer knew to expect tiny yields compared to commercial fields. Where a good field now might produce 250 bushels to the acre for commercial corn, you'd be lucky to get 60 to 80 bushels to the acre for seed corn, so the company compensates farmers for what the seed corn would have produced if it were planted to commercial corn and a top yielder. "You have these PBY fields, they call them. PBY is a 'production-based yield,'" Rick said. "And that sets the mark for yields for the seed corn fields, so you choose your very best ground and put your very best hybrid with your very best fertilizers, and all the other producers do the same. Those that are chosen then, in the fall, they go in and they harvest those. That gets your base yield for the year. They pay you for the seed corn fields at the rate that your PBY fields yield. The last couple years it's been right around two hundred fifty bushels."

On top of that base yield price, DuPont Pioneer also offers a premium for what they call the Big Dog Award. They compare the yields of groups of seed corn producers, farmers in a certain geographical area, what they call a "block," who planted similar varieties around the same time, and they rate them. "Say you were in the middle of the pack in a block of six," Rick said. "Then every acre, male and female, would be counted at 250—just average. But if you won out of those six, if your field did the best, then you're the Big Dog and would get some premiums. It's an incentive to not just plant the seed corn but to do your best to make the field produce."

Later, when Kyle was in the shop removing counterweights from the tractor after the end of planting, I told Meghan that I couldn't believe all of the know-how that goes into farming. Kyle had chosen the balance of what to plant, the varieties to plant and where. He had chosen how much fertilizer, pesticides, and herbicides to apply and when to apply them. And going forward, as the two of them took over the operation, they were going to be in a position of trying to amass their own land and build their own operation. They were constantly racing to get ahead, but it never seemed to be enough.

"There's a lot of things wrong with farming now, and the trend is more, more, more," Meghan said. Her hair was pulled back into a long braid, her eyes wide with emotion. "More chemicals, more herbicides, more pesticides. And more land. We need more—more, more, more. This land has maintained five generations, and I'm number six. We don't need more. We can get by on less." Then, suddenly, her face flushed with annoyance. "But we've also gone down this road quite a ways and the people who say, 'This is all a brand-new development and it's easily reversed,' don't know what the fuck they're talking about. You know?"

I nodded. "I know it's sensitive," I said, "that a lot of people ask why we can't just go back to raising organics."

"Because the emphasis on corn and soybeans started in the twenties. It's not a development just of the last few years. Yes, it's gone crazy since the seventies, but we spent a century going down this road and everything—the development of tractors, the development of seeds, the development of chemicals—everything has been built around a certain way of doing things. On the one hand, if we take a look at it and say, 'This is not the way we should be doing things,' then, sure, we need to do things different. But on the other hand, I also don't think that we should say, 'We should all start raising organics and we'll all just put on our overalls and we'll get out there and do it ourselves.'"

She shook her head at the very thought of it.

"Are *you* ready to go raise your own food?"

THE POKER RUN

June 2015

Meghan took a deep breath, steadying her nerves, as she rounded the corner and walked onto Green Street, the main drag through Clarks, Nebraska. Everything was decked out like it was the Fourth of July—American flags hanging from storefronts and public buildings, miniature flags stuck into flower planters and front yards. And on either side of the wide, brick thoroughfare were motorcycles and muscle cars, all lined up for the ninth annual Lance Corporal Brent Zoucha Memorial Poker Run and Dance. Meghan waved and gave hugs as she made her way through the crowd outside the VFW hall, past people sitting on a bench engraved with Brent's name and the date he was killed in action, past a group of women in fluorescent yellow t-shirts emblazoned with the words RIDE TO REMEMBER.

Inside, it was loud with people crowded around the bar

sharing pitchers of beer and getting officially entered into the Poker Run. As riders signed in, they were given the first card of their poker hand. From there, they would go on to the towns of Aurora, Harvard, McCool Junction, and Polk, drinking and getting a new card in each town. At the end, everyone would ride back to Clarks, compare poker hands to determine a winner, and then have a cookout and cap the evening off with dancing to a live band. Meghan found Brent's brother to give him a hug and then sought out Brent's mother, Rita. Rita was planning to go to the edge of town to see everyone off, but rather than riding along, she was going to head back home to let Meghan have some quiet time in Brent's old room, which Rita still maintained as a shrine to her son.

Precisely at 2:30 P.M., one of the organizers shouted that it was "clutches out." Everyone filed through the narrow door, fanning out to their bikes. Then they zigzagged down Green Street, their engines grumbling and handlebars weaving as they rolled in low gear like pacing tigers. At the edge of the downtown, the Patriot Riders hung a left and crept down Millard Street, its rows of American flags marking the path, as it had nearly a decade before. One by one, the riders coasted to a stop at the city limits, where they were thanked personally by Rita. The sun was high overhead, the temperature well into the 90s, but she wanted to greet them and make sure they knew what this meant to her. She hugged those on motorcycles and shook the hands of those in trucks and cars. One by one, they roared out onto the highway and she would wave down the next rider.

When the last had disappeared, Meghan went with Rita to her small house on the other side of town, on what was now named Brent Zoucha Lane, the place where Meghan used to go to hang out with Brent in high school, watching him play video games and smoke cigarettes. Rita unlocked the side door and chatted briefly with Meghan, but she never left the driveway. She had become an expert in the mechanics of grieving, and she seemed determined to let Meghan have her time alone.

As Rita drove off, Meghan turned to me. "It's okay," she said, ushering me through the doorway.

Inside, it was dark and icy with air-conditioning.

"Wow," Meghan said. "It's so much cooler in here, right?" She walked around the kitchen and the living room a bit, just taking in the place. She studied the recent newspaper clippings on the refrigerator, checked out the new family photos, framed and hung up around the house. At some point, it dawned on me that she was putting off going into the room with all of Brent's things. To let her know that I wasn't trying to rush her, I ventured a single question.

"Were you in this house often?"

"Spent a lot of time here," she said, looking over every inch some more. "As much time as I could. My parents didn't want us dating. We probably started freshman summer, sophomore year. I was pretty young, and he was full of it, super ornery."

She took a breath again. "Okay," she said, "let's do this." She fumbled through the keys Rita had given her, until she found the right one. She unlocked and then opened the French doors into the room.

Inside, it was hot, cut off from the air-conditioning of the rest of the house. Meghan flipped the switch, and the track lighting suddenly bathed the room in bright light. The walls and floor were covered with memorabilia—Brent's high school trophies, photos of him with Meghan at the prom, a picture of him after he enlisted. In death, every piece of him had become significant to Rita. One shelf held the contents of his shower kit sent home from Iraq: the last bar of soap he ever used, the last stick of deodorant and tube of tooth-paste. Whatever scrap of his life remained, it was on display here, but there were also items sent to Rita after Brent was killed. There were letters from senators and the governor of Nebraska and all kinds of tributes—drawings of Brent and paintings of him by people who had never even met him, just people, crazy or caring was hard to say, who wanted to make a connection with a parent who had lost her son to war.

And in the middle of the room, a lectern, positioned under a bright overhead light, held a 6-inch-thick photo album of still more memories.

"This is the scrapbook that she made," Meghan said. "She made one for all of us."

She flipped open to a series of photos of Brent in his bas-ketball and track uniforms. At 6 feet 4, he had been the star of the High Plains High School basketball team and was selected to play in the Central Community College All-Star game in 2005. In track, he took second place in both the high jump and the 400-meter relay at the Class D state track meet. Everything was there in the room: the trophies, the certificates, the med-als, the clippings reporting the results, all carefully arranged.

Meghan felt deeper into the scrapbook, turning several pages at a time. "This is Gunnery Sergeant Philips. He's the recruitment officer that signed him up," she said, then shook her head at Brent's serious expression. "He would never smile." Meghan flipped pages again and found an article from *USA Today*, quoting a family friend saying, "Second place just wasn't an option. I believe this drive to succeed is what led him to the Marines." She turned over more pages, and the scrapbook fell open to the calendar page for June 2006. Torn from Rita's wall calendar, the date for the ninth was circled and marked, "Brent's accident."

"Accident," I said.

"I think it's hard for Rita to talk about it directly," Meghan offered.

"Did you ever find out what happened?"

"Not exactly," she said. All the family had been told was that Brent had been part of a counterterrorism team operating along the Syrian border. While on patrol, one of the Humvees—whether it was Brent's or not was unclear—hit an IED. They radioed for support, and the second wave of Marines detonated still more IEDs. Regardless of the sequence of events, the Humvee that Brent was riding in had been hit squarely, killing him along with two of his battle buddies, Lance Corporal Salvador Guerrero, a mortarman from Whittier, California, and Navy Hospitalman Zachary M. Alday, a corpsman from Donaldson, Georgia. The three had been close; the official Marine memorial said that they "ate, lived and worked together on a daily basis," but Meghan didn't know anything about the other two killed in the blast.

"You know, I went off to Ireland, pretty much right after Brent's funeral," Meghan said. "I needed to get as far away as possible, and I think that trip saved me. But I also spent a lot of days over there searching online for stories about what had happened and any details of the guys he had been with."

One article featured a photograph of Sal Guerrero's mother stretched in desperation across her son's casket, as if she were begging that it not be lowered into the ground. Meghan left her email address, phone number, and home address in the comments section, hoping Sal's family would see it and write to her. "I just wanted to get in touch with your family and find some things out," she wrote, but Sal's family was still in deep shock. Not wanting to worry his mother, he'd never even told her that he'd been deployed to Iraq; when he called from the forward operating base, he always told her he was on a training mission in Japan and would be home soon. Eventually, Meghan made contact with Laura Almanza, Sal's girlfriend, who had known where he was all along.

Meghan flipped another group of pages in the scrapbook and landed on the program for Brent's memorial at the base in Twentynine Palms, California. "This is when we finally met," Meghan said. "I decided to fly out with Rita, just to be there for her, and Laura came over from Whittier." After the ceremony, some of Brent's friends came up to introduce themselves, but they didn't talk about what happened and Meghan didn't ask. "I was just a girlfriend. I wasn't a wife, and I wasn't family. I think that's why Laura and I bonded right away. We were in the same boat." When the reception was over, they all went out on the town to blow off steam. "I was only eighteen, so I may or may not have had way too much Patrón."

At some point, they all decided to get tattoos. They spilled out onto the streets of Twentynine Palms, not knowing exactly where they were going. "But in a town with a Marine base," Meghan said, "a tattoo parlor is never too far away." When they found a place, Laura decided right away to get a cross on her back with a rose. Rita settled on a purple heart on her chest. Meghan still wasn't sold on the idea and sat paging through the sample booklet, when she landed on a tattoo with Brent's name. His friends had already been in. "So I decided to do it," Meghan said. "I asked for the Marine Corps motto— Semper Fi—along with Brent's jersey number in basketball."

She laughed and pulled up her pant leg, revealing the tattoo on her ankle. "The guy drew it freehand, but it didn't come out too bad."

She closed up the scrapbook and looked up into the lights. "You know, I don't know what would have happened between Brent and me. I'll never know." In recent weeks, Kyle had finally worked up the courage to formally propose. Meghan was wearing the engagement ring that he had given her. Rita was Meghan's first phone call. It was hard, but the last thing she wanted was for Rita to find out from someone else. "She just said how much she likes Kyle. She wants to see my wedding dress and all that stuff. I know that none of this has been easy for her."

Before switching off the lights and locking up the room, Meghan took one last look around. There was nothing new—and never would be—but she studied the photographs of Brent, especially those of the two of them together, as if she were looking for some hint of the tragedy ahead. But there was nothing there but two kids, young and having fun. Gussied up for the prom. Goofing around at the driving range. Hanging out together on the same couch that sat just outside the room—Meghan perched on Brent's lap and laughing, his arms slung around her.

"That was the last time we were together," she said, "around Christmas, right before he left."

PART FOUR

IRRIGATION NATION

Summer 2015

Rick turned a yellow dial until it locked into place with a hollow clank. The royal blue control panel, emblazoned with the logo for Valley Irrigation, hummed to life, and across the windswept field, a light atop the enormous center-pivot irrigation system started to blink. "That means everything is on," he said. The metal behemoth of the pivot stretched a quarter mile from end to end. Its humped spine and rib-like trusses, girding an array of dangling hoses and sprinkler heads, looked like the skeleton of some kind of robot *Brontosaurus*, poised to waken from its prehistoric slumber. Rick tipped his cowboy hat back with his thumb to get a closer look at the digital display. He checked the speed and pump pressure, to make sure the wellhead was still off. Seeing that everything looked set, he gave me a little shrug. "Then you start it walking," he said. With the push of a button, the pivot eased forward. The twinned drive wheels under each triangular tower suddenly lumbered to life, creeping clockwise along its circular path.

It was still June, early in the season, and Rick hadn't yet planted this final field of soybeans. Before the crops were high and the rows were wet with rain and irrigation water brought up from the pivot's central well, he wanted to be sure that everything was in perfect running order. "Every year, we go through our pivots before planting and drain any water accumulation in the gear boxes. Each has an electric motor going to two drive shafts, so we do all that greasing; we don't want those seizing up. And then, we just visually inspect the whole thing. Do all tire pressure. If any are suspect—so weather-checked that they're not going to make it a season—we usually just go ahead and replace them. I can tell you what ruins a Sunday afternoon, and that is taking one of those sons of guns—we're talking about a chest-high tire—and shoving it through a planted field on a muggy, hundred-degree day. And you have to carry your jacks and your blocks and your spinner wrench. You want to talk about a workout."

We climbed into the truck and drove across the field to where the pivot was walking in a clockwise direction. "You want to be sure to park on the left side of it," Rick said with a grin. "Meghan had one time where she parked a seed cart in the wrong spot. It could have tipped over, but luckily the tower rolled up to the cart just right and idled there long enough that it just switched off." Rick stepped over the rows, inspecting each span as he talked, checking that the gear boxes looked like they were turning and the towers were moving in alignment. When he seemed satisfied that all was in working order, we drove back to the control panel. Now he switched on the water pump to double-check that it was drawing properly from the well.

"What we're shooting for on this level of ground is about an inch," Rick said over the roar of the pressurizing wellhead. "It won't run off with an inch. That takes approximately three and a half days." He said that, in a good growing season, he hoped to put just three to four inches of water on this whole 160-acre quarter section, most of it in a single spate in July. That doesn't sound like very much, but it can be the difference between a bumper crop and getting wiped out for a whole year. When Heidi's grandfather first dug this wellhead in the late 1940s, irrigation was still in its infancy. In those days, a pump would bring water to the surface and dump it into massive catchment ponds, which could be released into open furrows during dry times. Known as "flood irrigation," it was a game-changer for farming. Using well water, instead of just relying on surface streams, made it possible for farms to spread away from river valleys, adding arable acres and allowing farmers to start planting crops in places they never would have dared before.

But the amount of water pumped and put onto a field using flood irrigation was hard to control. Catchment ponds were susceptible to evaporation. Flooding furrows caused widespread runoff, leading to loss of water and topsoil. That's why the development of center pivots in the 1950s, something as simple as applying water evenly over a field in rain-like droplets, rather than ground-level flooding of furrows, proved so revolutionary—and why it was embraced so quickly and so universally. Even American city-dwellers who have never been on a farm have looked out their airplane window and seen the green crop circles created by these irrigation systems. That patchwork quilt of lush crop production has become the unofficial symbol of our national agricultural know-how and is one of our main ag tech exports around the world.

The simple shift in irrigation strategy made it possible to raise crops in places with unreliable rainfall, expanding our potential for production. But the more precise delivery of groundwater—along with decades of refinements in water application technology—has also made it possible to conserve that precious resource, by applying only the necessary amount at the moment that it is most needed. "See those hoses," Rick said, pointing to the black rubber tubes dangling from the pivot's spans overhead. "When the corn is high, they'll hang right down to the tops of the tassels. It used to be that the old emitters, the sprinkler heads, had impact spray nozzles and were set way up high, so it would put out a wide spray pattern, a fine mist, but we lost so much water to evaporation. So now, they're down at leaf level, and, for efficiency, we use Nelson wobblers that throw large droplets. By applying drops, more like actual rain, more of the water application makes it to the roots, rather than sitting on the leaves and evaporating. We're trying to do as many things as we can to minimize usage."

But even reduced drafts from wells, applying just a few acre inches in a whole growing season, are drawing a lot more water than you might think. Rick led me over to the meter, attached to a pipe coming up from the well. He ripped up handfuls of tall grass that had overgrown the gauge. "Look at that," he said, pointing to the small lettering under the spinning numbers. "'Gallons *times one hundred.*' We're talking millions of gallons of water that get pumped from the Ogallala Aquifer." Sixty years after the advent of center-pivot technology and the movement toward water-intensive crops like corn, those millions of gallons were starting to add up, catching the eye of environmentalists and farmers alike.

A 2015 study of U.S. Geological Survey data compared the depths of more than 32,000 wells nationwide over the previous two decades. The results were alarming. Across the country, water levels had fallen in 64 percent of all wells, with an average decline of more than 10 feet. In the Ogallala Aquifer system, which supplies groundwater for crop irrigation not only in Nebraska but to eight states from Texas to South Dakota, the declines were especially pronounced. In much of southwestern Kansas, wells were down to 25 percent of the water that existed when the aquifer was first tapped less than seventy years before. In the southern High Plains of Texas, near the edge of the Ogallala, water levels had fallen more than 100 feet in places, leaving many farmers without any water at all. In 2011, the U.S. Department of Agriculture announced that the government would be investing over $70 million in the Ogallala Aquifer region, "to help farmers and ranchers conserve billions of gallons of water." But the Great Plains was soon plunged into a multi-year drought, and, instead of declining, water usage shot up dramatically.

After he was done checking over the pump and had twisted the yellow dial of the control panel back to its off position, silencing the humming system, Rick seemed caught for a moment in the calm. He confided that he worried sometimes about how much water industrial ag used on the Great Plains. "You know, if I've got one issue that I really care about," he said, "the environment would be it." He was proud that Nebraska's unique system of Natural Resource Districts, governing responsible water use at the local level, had become a model of management and stringent enforcement of consumption, but even in eastern Nebraska where the aquifer recharges faster than any other part of the system, the levels had still dropped noticeably in Rick's lifetime. "I have seen nature repair itself, but it takes time," he said. "If humans don't give it a chance, we'll end up stressing the resource, and I think it could be bad."

I was struck by the turn of phrase. "Stress the resource," I said. "You mean we could use up all of the water?"

Rick hesitated, then waved me back toward the truck. We climbed back into the quiet of the cab, out of the wind. "Look," he said finally. "If rainfall patterns keep shifting, we keep getting hotter summers and less rain, and people don't conserve the groundwater, I'm not saying there will be no more water. But there won't be enough to raise crops, not like we've been doing. There will just be grass in Nebraska."

Rick promised to take me out west to where he grew up in Curtis, maybe at the end of the summer when it was time to bring the cow-calf pairs back from pasture. Not far from there, he said, at a place called Dancing Leaf Earth Lodge, archeological excavations had uncovered the remains of a once-thriving village, part of the Upper Republican Culture, an ancient precursor to the Pawnee. Digs suggested that those people, after centuries of farming corn, were forced to abandon their homes in the late fifteenth century, displaced by a catastrophic drought that lasted generations. "We brag about toughing it through years of hard times," he said. "Sure, we've made it through droughts. Three-, five-, seven-year droughts. Terrible, but we got through. But what about the fifty- or hundred-year droughts?" The sites of many of those abandoned villages had been found buried under a thick shroud of dust—proof of dry times and roiling winds that rendered the area uninhabitable for multiple lifetimes.

"They left and never came back. They don't know for sure what happened to them. Could it happen across the Great Plains today?" Rick asked. "Absolutely."

FINDING WATER to grow food has been the central challenge of life on the Great Plains since the earliest days of white settlement. When the United States, in its zeal to displace Native Americans and settle the Free Soil debate with the Confederacy, passed the first Homestead Act in 1862, government surveyors fanned out to plat a patchwork of one-mile squares, which were then subdivided into quarters of 160 acres each. A "quarter-section" was awarded any free citizen of the United States, man or woman, native-born or immigrant, with just one major catch: you had to live on the land and farm it for at least five years. This not only meant concocting some way of building a home on the treeless prairie, but also finding enough fresh water to sustain crops.

On the semi-arid plains of Nebraska and neighboring states, where surface streams often ran dry during the summer months—right when water for irrigation was most needed—farmers had little choice but to dig wells. In some places, large communal wells were undertaken to serve whole communities. In 1887, for example, a group of determined Kansas farmers, using nothing more than shovels and picks, dug a stone-braced well to a depth of over 100 feet to supply freshwater to the town of Greensburg. But more often, the wells were shallow and hand-dug. Farmers erected windmills to pump small amounts of groundwater and divert it to the wide furrows of their fields or into catchment ponds. Life on the hardscrabble plains was arduous, and often hand-to-mouth. Nevertheless, by 1890, the Homestead Act had settled some 2 million people on nearly 375,000 farms.

But in the drought years that followed, just as cities like St. Louis, Omaha, and Denver were turning into bustling metropolises built on income from livestock and grain exchanges, the entire Great Plains region was devastated by blistering heat. It wasn't just an unlucky few who were pushed to the brink of bankruptcy. By the harvest of 1894, one newspaper reported, nearly 38 percent of acres planted with corn in the middle states were either destroyed or abandoned. In Nebraska, where high temperatures were accompanied by scorching winds, the smell of parched corn filled the air. Meghan's great-grandfather, Harry Harrington, may have struggled through those years and eventually lost his grip on the family land, but many more simply packed up and left, making their escape, according to another paper, "while they had something to do it with." The drought revealed that raising enough grain for a farm family to thrive would require greater land allocations than the Homestead Act had originally provided—and a lot more water.

In western Nebraska, where the drought was worst, a group of locals formed the Farmers Irrigation Project, expanding an existing small-scale canal that had started diverting water out of the North Platte River right along the Wyoming border, near Henry, Nebraska, in 1887. My grandfather eventually worked on the channel that resulted from that project, the Tri-State Irrigation Canal—a network of open waterways that stretched from the foothills of the Wyoming Rockies, across the northeast corner of Colorado, and into western Nebraska. He was a ditch rider in the Wildcat Hills, doling out water allotments, for more than thirty years, so on one trip out to that part of the state, I'd stopped in to see Kevin Adams, who had worked his way up from ditch rider to construction engineer to supervisor of the Minatare division to assistant manager to his current position as general manager of the whole district. Though still in his fifties, Kevin had worked for Farmers Irrigation for more than a quarter of the lifetime of the canal. So when I asked what changes he had seen in his years on the job, he broke out in a crooked smile.

"Nobody rides ditch with a horse anymore," he said.

We were sitting in his sprawling, wood-paneled office just off the main drag in Scottsbluff. "No," he said, "I never saw that. But this system is old enough that everything was built with horses and slips. They didn't have dredging machines or anything. I've got an old ditch rider's manual in my desk that says you have to supply your own horse, your own feed . . ."

He paused a moment and smiled again. "Do you want to hear all that history?"

I nodded for him to continue.

"So, this system has a water right of 1887—when the first irrigation supplies were diverted out of the North Platte River," he began. "By 1891, they had ten miles of small canals built that irrigated about thirty-five hundred acres with the natural flow of water. But they soon found out that, since this part of Nebraska isn't on an aquifer, they couldn't sustain crops through the summer without storage water." The project was suspended—but, in 1902, with passage of the Reclamation Act, the federal government agreed to bankroll a dam in Wyoming that promised to deliver enough water for 150,000 acres, by a new canal dubbed the Pathfinder. At the same time, Heyward G. Leavitt, an entrepreneur who believed that with proper irrigation sugar beets could thrive in the Platte Valley's rich soil, bought out the remaining interests in the Farmers Irrigation Project—in return for delivering free water in perpetuity to the land of the original owners. Leavitt got a supplemental storage contract with the Pathfinder Reservoir and purchased 36,000 acres, where he extended the old canal and taught tenant farmers to raise sugar beets.

Within a decade, Leavitt's Tri-State Canal was completed—and the sugar beet had transformed the valley. The beets themselves were reliable and profitable. The Great Western

Sugar Company provided seasonal factory jobs, first in Scottsbluff and then at a second plant in nearby Bayard. The byproducts of sugar refining (both beet tops and leftover beet pulp)
could be processed into feed for cattle, stabilizing the beef
industry and creating jobs on ranches and bringing packing-
houses into towns along the railroad. And the manure from
the ranches could be used to enrich irrigated soil to grow a
greater diversity of crops—most notably soybeans, alfalfa, and
eventually corn.

"At that time," Adams said, "everything was open
channel—seventy-five miles of a main canal, which we still
have today; probably 288 miles of open laterals. And, of
course, we maintained most of the creeks and drains in this
area—about ninety miles of that, too." All told, the Tri-State's
storage supply in Wyoming's Pathfinder reservoir is capable
of delivering 1.1 million acre-feet of water, but even in the
early years, the river couldn't deliver enough to meet all of the
agricultural demands of the growing diversion projects. By
midsummer of each year, the Platte River in the central part
of the state and points farther east always ran dry.

So at the same time that the Farmer's Irrigation District
was digging canals, the U.S. Geological Survey had commissioned a study of groundwater resources on the Great Plains.
In 1897, working at a spot west of Ogallala, Nebraska, government geologist N. H. Darton surmised that the abundant
wells in the southwest corner of the state were drawing from
a great storehouse of ancient water held in place by an enormous underlying layer of limestone. "Extending from Kansas
and Colorado far into Nebraska," Darton wrote, "there is a
calcareous formation of late Tertiary age to which I wish to
apply the distinctive name Ogalalla formation."

Darton described the Ogallala as a kind of enormous, untapped reservoir, like a lost freshwater lake waiting through the millennia to one day emerge and regreen the plains. Over time, however, geologists have come to understand the aquifer as something much more complex—and difficult to manage. Ann Bleed, a former state hydrologist and director of the Nebraska Department of Natural Resources, explained to me that the Ogallala isn't even a single aquifer but many aquifers, studied and managed as independent parts of the High Plains Aquifer System.

When the system began to form over 100 million years ago with the uplifting of the Rocky Mountains, the shallow inland sea that had covered much of North America slowly receded, and rain and snow from the rising craggy peaks ran off, forming rivers that raced east from the mountains to the flatlands. In time, the braided channels filled with silt and gravel, creating spaces that trapped a vast basin of freshwater commingled with the alluvial soil. Contrary to popular views, however, an aquifer is not a giant underground lake, with all that implies. It is not a round-bottomed bowl filled with water, but rather the flooded and filled in remains of valleys and canyon walls, more akin to the Badlands than the flat prairie we see today.

So in places where rivers once ran, the Ogallala can be quite deep, but in areas where the bedrock is the high top of an ancient butte, it may be shallow and cut off from other parts of the system. Some parts of the aquifer, like those in eastern and central Nebraska, can replenish, or "recharge," rather quickly, because, by the good fortune of geology, the aquifer there is connected with the substrate from the Rocky Mountains. The underlying bedrock still tilts to the east, carrying underground water resources in that direction. And the area around York and Hamilton counties also benefits from water slowly percolating out of the Sandhills to the north and west. In Nebraska, the aquifer can seem like an inexhaustible resource, but, in many other places, wells can be inadvertently tapped out—and take hundreds of years to replenish. Still other spots, like the aquifers underlying western Kansas and wide swaths of the Texas Panhandle, refill so slowly it's as if they don't recharge at all; geologists call their water "fossil water," because it is more akin to oil, a finite and nonrenewable resource. But no one knew that at the time they were discovered.

In the 1870s, interstate railroads brought ranchers to West Texas, where they could run cattle on flat expanses of public rangeland. At the turn of the twentieth century, when cattle baron Christopher Columbus Slaughter, the, godfather of the Texas Longhorn, began setting wells for his growing herd of 20,000 breed stock, most of his digging crews didn't even have to drill for water. They simply used a horse-drawn trench digger to deepen a natural arroyo or a buffalo wallow, which would fill with water whenever rains fell on the already saturated soil. In other places, Slaughter hired men to install windmills to pump water into stock tanks spaced five miles apart, but the water was so close to the surface that they could set a well with nothing more than shovels.

Thanks to readily abundant groundwater, the economy of West Texas, especially in cattle towns like Plainview, boomed. The downtown there saw the construction of the Granada and Fair vaudeville halls. The city erected the 3,000-seat Plainview Municipal Auditorium. In 1929, Hilton built a seven-story hotel with 125 rooms, all with private baths. Seeing the success of irrigated cotton fields, other farmers were encouraged to start planting wheat and feed crops. B. F. Yearwood, a local entrepreneur, erected a 50,000-bushel corn mill at the edge of town. The Harvest Queen flour mill, which opened its doors in 1907, was completely rebuilt with soaring concrete grain elevators. The city was so convinced of its limitless water supply that it planted more than 50,000 shade trees as part of a beautification project. In 1937, at the height of the Dust Bowl, the McCormick company issued a postcard, intended to attract drought-stricken farmers from Oklahoma. A young farm girl is shown holding the lead of a beef cow and standing next to an open irrigation well, its windmill spilling water into a verdant-ringed stock pond. The legend reads: "This modern farmerette has no rainfall worries."

Farmers in eastern Nebraska weren't so lucky. Though groundwater was abundant there, it was not as near to the surface as in West Texas. Irrigation wells were reliable and sustainable if they were dug along the banks of the Platte River where the water was closer to the surface, but farmers there already had the region's most abundant water supply and were often reluctant to shoulder the cost of drilling and putting in a wellhead, or the price of buying and maintaining oil-powered pumps. Some employed wind power, but a single windmill could only siphon enough water to irrigate five acres or provide for thirty cattle—hardly enough to get farmers through the dry times. In 1928, the Nebraska Agricultural Extension Service lamented that "the underground water supply is abundant," but there were insufficient means of "lifting it to the surface and applying it to the land." Toward the end of the 1920s, as prices of corn and wheat sagged due to over-supply worldwide, there was even less incentive to invest in irrigation equipment. Nebraska farmers were content to keep their input costs low and get by on the twenty-five-bushel-per-acre yields they had steadily produced since recovering from the crisis of the 1890s.

Before long, however, the middle of the country was hit with the drought against which all other droughts would come to be measured. In 1934, the start of the Dust Bowl years, Nebraska's statewide average per-acre yield of corn was barely two bushels—and didn't rise above ten bushels per acre again for another five years when Henry Wallace's drought-resistant corn hybrids brought the country back from the brink of starvation. But then, even before the country could stand on its feet, we were drawn into World War II and five years of food rationing.

During those early years of the war, when the fighting in Europe was still a pitched struggle between heavily mech-anized armies, tractor manufacturers like John Deere and International Harvester boomed. Government contracts meant the assembly lines ran day and night, building engines for heavy trucks and tanks. My grandfather, driven out of Nebraska by the drought, moved his wife and infant son, my father, to East Peoria, Illinois. There he worked at the Cat-erpillar plant, making tracks for tanks as well as industrial scrapers used to build makeshift landing strips for supply planes and bulldozers to clear roadways for trucks to carry those supplies to the front.

After the war, all of those companies shifted resources to developing peacetime applications for their heavy equipment, including diesel-powered pumps that could finally convey all of that groundwater underlying much of the American bread-basket to the surface where it could be used for crop production. After a long decade of going hungry, it's little wonder that Americans were collectively determined to find a way to ensure that no water shortage would ever again mean going without.

🌿

In the late 1940s, while tenant farming in Colorado, a farmer and tinkerer named Frank Zybach attended the Irrigation Field Day held near Prospect Valley. He watched the demonstration crew set the pipes on an irrigation system, then tromp through the mud to move them. The arrival of abundant water had been a godsend to Colorado and Zybach's native Nebraska, but the early irrigation systems that delivered that water to the field were little more than a series of bulky metal pipes that had to be lugged into place and then reset two or three times a day. It was a mucky and painstaking job, and not a terribly efficient one.

Zybach later remembered thinking, "There has to be a better way." He began to develop and eventually patented a self-propelled system, in which the force of water flowing through the irrigation tube would slowly turn the drive train of supporting wheels. By anchoring the pipe to a central wellhead, the system turned in a perfect circle, watering the field evenly without any work by the farmer. Zybach's system became known as "center-pivot irrigation." In 1952, he moved back to his hometown of Columbus, Nebraska, and partnered with local equipment manufacturer A. E. Trowbridge on building and marketing the systems.

Despite their ingenious design, early center pivots were plagued by malfunctions and were a nightmare to maintain. The long central irrigation pipe was prone to leakage, which

would drop the water pressure enough so that the whole system ground to a halt—often flooding a single spot in the field
in the process. Farmers, leery of the expense and hassle of
owning an untested piece of equipment, were reluctant to
buy. In the first two years of production, Zybach later told
the *Lincoln Journal Star*, he and Trowbridge sold just nineteen
center pivots. After several years of struggling, they decided
to license the technology to Robert B. Daugherty, the young
owner of a small farm equipment company called Valley
Manufacturing, not far from the banks of the Platte in Valley, Nebraska. Zybach hoped that Daugherty could improve
the design and make the system profitable.

Daugherty spent months and months working on a pipe-
making machine that could take a flat piece of steel, roll it into
a wide-gauge tube, and then automatically weld it shut. When
he had finally perfected the process, he was eager to show it
off. In 1953, one of the first buyers of a center pivot from Valley was a young man recently returned from the Korean War
named Walt Sehnert. "One day my dad and I stopped in at the
Valley plant to see this new pipe-making machine," Walt told
me. They were greeted by Daugherty and given the grand
tour, culminating in a demonstration. "Daugherty was a dap-
per little man, ramrod straight," Walt said. "He even carried
a little swagger stick, favored by some Marine Corps officers.
As we walked through the plant, he explained the various
phases of manufacture that we were seeing, and slapped the
side of his leg with this stick from time to time." When they
finally reached the pipe-making machine, it didn't seem to be
working right. Daugherty approached the two men who were
tinkering with the machine and offered a couple of sugges-
tions. "Then he came back to where we were standing. He

said nothing, but clenched his teeth, turned red, and glared at the machine, beating a tattoo with that swagger stick against his leg." A minute later, one of the men gave the thumbs-up, and the machine rolled out a perfect 6-inch pipe.

Sehnert bought and installed the system. Soon other farmers followed suit—and just in the nick of time. In summer of 1953, furnace-like temperatures spread across the Central Plains, pushing hot, dry air across the American breadbasket. The drought that followed held the region relentlessly in its grip. For nearly four years, the middle of the country, from the panhandle of Texas all the way to the Sandhills of Nebraska, experienced low rainfall and stretches where the mercury topped 100 degrees for weeks at a time. The federal government spent $3.3 billion on assistance to farmers, and the beef industry in Texas was nearly destroyed in what became known as the Great Cattle Bust.

But the lucky landowners of eastern Nebraska who had invested in center pivots were able to weather through the drought. Other pivot manufacturers soon popped up around eastern Nebraska—Zimmatic in Lindsay, Reinke in Deshler, and Olson Brothers in Atkinson—and they began setting up dealerships across the Great Plains. In the decades that followed, farmers from North Dakota to Texas turned to center-pivot irrigation to provide extra water during key growing times and help them through dry spells.

By the end of the 1970s, there were more than 18,000 center pivots operating in Nebraska alone and some 30,000 in other parts of the Great Plains, altogether irrigating more than 20 million acres. The rapid expansion of center-pivot technology allowed farmers to plant on more and more marginal land and to venture into water-intensive crops, leading to dense planting of corn. In just a few short decades, the arid plains were transformed. The changes were so profound that astronauts aboard Skylab reported seeing a patchwork of green circles stretching for miles across north-central Nebraska.

"Passengers on commercial jet airliners increasingly notice the same sight," wrote William E. Splinter, chair of the Department of Agricultural Engineering at the University of Nebraska, in an article for *Scientific American*. "What is being observed is perhaps the most significant mechanical innovation in agriculture since the replacement of draft animals by the tractor."

Today, a tower from one of the original center-pivot systems manufactured by that revolutionary pipe-making machine is still on display in the lobby at Valley Irrigation's manufacturing headquarters. Matt Ondrejko, vice president for global marketing, told me that the exhibit honors Valley's central role in averting agricultural disaster in the 1950s and building the large-scale ag economy that followed. Touring the sprawling complex of factories and machine shops bounded by test fields, it's clear that the folks at Valley recognize how much the future of the company depends on continuing to innovate, particularly in the area of water conservation.

"We can't run a business selling irrigation systems if there's no groundwater for irrigation," Ondrejko said. The company is able to tackle the challenge by closely monitoring every step of the construction of its center-pivot units—from the hand assembly of the plastic sprinkler nozzles to a massive crane system that lowers spans of pipe into a gargantuan zinc-and-nickel bath.

In recent years, Valley has focused its research and

design improvements on two main areas: the sprinkler heads and the control mechanisms. By collecting field-moisture data and contrasting yield returns in comparable fields irrigated with different sprinkler heads, engineers are able to refine the spray patterns and the drop patterns, to reduce loss through evaporation. The latest design innovations have focused less on fine droplet dispersal and more on getting water all the way to ground level. After all, the goal is watering root systems, not leaving droplets on leaf surfaces where they will be lost. "There's really precise science in how big the drop is," Ondrejko said, and "how close we get it to the ground."

Valley has also invested in developing variable-rate irrigation, which allows individual sprinkler heads to be turned on and off in particular parts of the span as it passes over different sections of the field. The objective, Ondrejko said, is being able "to precisely spoon-feed the crop—when it needs it, where it needs it, no more, no less." Perhaps most important is the development of exact application monitoring, so that farmers can collect precise information on how many acre-inches they are using and where. That's where technology meets water management efforts in the field.

After decades of pumping their groundwater to keep up with farmers in other parts of the Great Plains, the wells in West Texas are running dry. In Plainview, the wide, brick-lined avenues are sun-scorched and shuttered now. The plywood bolted over the front windows of the Hilton is painted with a mural of the former lobby, but the locals now speak only of ghosts lurking in the abandoned upper rooms. As the community limped through the droughts of the fifties and seventies, and the earth-parching heat wave of the early eighties, the cotton industry struggled to stay profitable, and gradually the area returned more and more to cattle feeding and beef processing. With the arrival of the Great Recession and the onset of the drought in 2011, Plainview seemed to lose what was left of its other industries: the White Energy ethanol plant closed temporarily, the Peanut Corporation of America plant shuttered due to a salmonella outbreak, and the Harvest Queen Mill & Elevator (by then owned by Archer Daniels Midland), also closed—collectively taking with them more than 1,000 jobs in a town of just 22,000 people.

To avoid a similar fate in Nebraska, David Eigenberg, who heads the Upper Big Blue Natural Resource District, the same NRD that monitors Rick's wells, told me that the key is more than just better equipment. Technology has led the way to "improved irrigation efficiency," but in many cases, that has only encouraged farmers to apply water to their fields more often. "You might have the Cadillac out there," he said, a center pivot with variable-rate application and zone sensing to automatically apply only in places where water is most needed, but all that advanced technology is useless "if you don't understand the management tool."

And that's Nebraska's real advantage. The state is broken into twenty-three Natural Resources Districts, which were created in 1972 to manage water at the individual watershed level. The NRDs apply a doctrine of "reasonable use" of groundwater, as determined and overseen by locals elected to the NRD boards. By contrast, Texas adheres to a doctrine called "rule of capture," which essentially allows farmers to use whatever groundwater they can tap from their property. As farmers come to see the resource as finite, many may actually increase their usage, in order to keep up with perceived competition from their neighbors—a phenomenon known as "the tragedy of the commons." To really reduce water usage, you have to change a culture that views irrigation as an individual right, rather than a method of accessing a shared resource. In the end, Eigenberg said, the change has to come from "the guy in the tractor seat."

To encourage that change, the NRDs have engaged in a range of tactics. Some have started requiring metering on new wells, which has helped farmers to track field-by-field usage and encouraged them to reduce their draft. Some have offered education courses or water-use counseling to help farmers see how overdrafting is costly in terms of the power used on the pivot pump and often in top-soil loss due to run-off. And some, like the Upper Big Blue NRD have undertaken even more ambitious initiatives.

The Upper Big Blue, which contains significant acreage classified as "high risk groundwater areas," launched a study in 2012 comparing crop yields in neighboring fields with identical center-pivot irrigation systems—but with one managed by the farmer and the other managed by the NRD. The farmer applied water according to his own judgment, while the NRD used soil-moisture data collected from a network of monitors inserted into the edges of the field. The NRD achieved a nearly identical yield—98 percent of what the farmer harvested—while drawing only a third of the water. The latest advancement is a wireless soil-moisture monitoring system that sends data to the farmer electronically. Future improvements will send that data directly to the center pivot, automatically turning the system on when irrigation is needed and applying water only in the necessary areas. It's hoped that these new technologies will be able to universally drop usage in these high-risk areas to the point that the aquifer might actually begin to recharge.

But even bigger, more sweeping changes may be needed. To address the problem adequately, we may need to rethink what kind of food we grow where, and how much agriculture is feasible in certain landscapes. The center pivot allowed the repopulation of many areas that had been vacated during the Dust Bowl—areas that simply couldn't sustain crops without a new source of water. That new technology, coupled with other advances in agriculture, put more than 100 million acres of marginal land into farm production, much of it for growing water-intensive crops, like corn and soybeans, which today are raised primarily for livestock feed and biofuel, not for human consumption. Farmers have also been encouraged to turn away from more drought-resistant cash crops— such as sugar beets in western Nebraska and sorghum in Oklahoma—in favor of commodity grains that are supported by higher farm subsidies and crop insurance.

The Ogallala water boom made it possible for the United States to become the world's granary, but now, as wells run dry in western Texas, Oklahoma, and Kansas, it may be time to start rethinking our usage. If we don't do something, Don Wilhite, founding director of both the National Drought Mitigation Center and the International Drought Information Center at the University of Nebraska–Lincoln, believes that the effects of this rapid drawdown could be catastrophic. In a 2013 report, written for the Institute of Agriculture and Natural Resources at the university, Wilhite warned that greenhouse gas emissions, if unchecked, will cause temperatures in Nebraska to increase by as much as 9 degrees Fahrenheit in less than a few years. That projected increase in temperatures means that summertime highs will regularly surpass 100 degrees by 2060.

Those extremes falling, as they would, during the peak of the growing season would create untenable demands on groundwater. Even if rainfall were to increase, it would not be enough to offset the loss of soil moisture caused by extreme heat. In a hotter climate, it will take more water to generate the same crop yield, even with genetically modified grain hybrids. Unless we change farming and water management, he explained, there simply won't be enough groundwater to combat such dry conditions.

Most important, Wilhite said that the loss of groundwater resources would set off a feedback loop with much broader effects. He explained that pumping all of that deep, cold water from the Ogallala and spreading it across many acres has artificially lowered air temperatures and increased humidity. That human-induced microclimate has masked the effects of climate change by forestalling climbing temperatures on a regional level. If we run out of that water, temperatures will rise sharply. Crippling drought will become the new norm, turning the Central Plains states into a permanent dust bowl.

Wilhite told me that his greatest worry is that farmers tend to brush off such dire predictions, insisting that they have lived through many hard times. They say that their grandfathers got through the Dirty Thirties and innovated their way out of the Great Cattle Bust in the 1950s. "Farmers say they're used to variability," Wilhite said, "but these projections are way outside the range of anything we've ever encountered."

On a sun-bright morning in Loup City, Nebraska, Rick walked around a maze-like complex of holding pens, inspecting thick-necked and hulking black Angus bulls. Their coats glistened like oil slicks in the daylight, and they stepped broodingly around their pens, like boxers before a title fight. The day's auction was set to begin just after lunch, and Rick was going over the program, checking each breeding male, one by one. All of the bulls for sale that day were from Dutch Detleff's ranch in Ravenna or Loren Treffer's place in Broken Bow. Rick knew Dutch from way back. He'd bought from him before, and he trusted the quality of Dutch's genetics and his instincts on what breeders call the "phenotype" of particular bulls—the visible characteristics that they read as indicators of the size and health of their future offspring. At one pen, he paused to show me the bull's number in the program along with the entry for his stats.

"EPD is your expected progeny differences," he said. "That's the estimate of how that bull's calves will perform over the average weight for the breed—at birth weight, wean weight, and yearling weight." The ideal was finding a bull that produced calves that were barely over average birth weight, to keep down the chance of losing cows during calving. "We had fifty heifers calve this spring," Rick said, "and didn't have to pull a single one." But once they're safely born, you want those calves to put on weight swiftly. Many of Dutch's bulls were expected to sire calves that were just a pound or two over standard birth weight but then would put on an extra fifty pounds by weaning and could be over a hundred pounds heavier than average by the time they were a year old. "With the price of beef over six dollars per pound," Rick said, "getting a good bull can add six hundred pounds to the sale price of each calf. Multiply that across a year's births, and it really adds up."

But bigger isn't always better, Rick said, especially when you're talking about cow calves. He pointed to another category on the grid, marked "Milk." "That tells you what their offspring cows will produce for milk once they're pregnant and lactating," Rick said. You want a cow that will produce enough milk to be sure that her calves are well fed and put on weight, but you also want to avoid genetics that produce udders that get too big. "An extra twenty-five pounds is about the top range," he said. "If you get too much milk, then you start to get bad bags. The calves can't nurse or the teats get mastitis, and you can't keep them in the herd for very long."

So the trick, as with everything on the farm, was factoring in your input costs and your relative risks. Calves that nurse aggressively will put on valuable pounds but require stouter cows and more feed for the mothers. Once weaned, those calves will need more food, too. Because Rick grazes his cattle almost exclusively on grass, it's not the cost of food that he's worried about so much as time and energy—moving the herd from one pasture to another, rotating weaned calves out to Curtis where they can bulk up on open rangeland and then bringing them back for sale. And you have to weigh the prospect of big bull calves against the risk of oversized cow calves. Factoring those costs and risks into the potential for increased profit, Rick settled on a total number he was willing to pay for two bulls that caught his eye. He was doubtful that he could get both for that price, but it was a number he wouldn't go past.

Inside the sale barn, the bidders were being called to the bleachers for the auction to begin. The auctioneer, who had driven in from Colorado for the day, sat high above the sale ring, leaning close into the microphone. He wore a tall, black Stetson and reading glasses, and he introduced the breeders in the buttery tone of a call-in radio host. From the moment the first bull was finally brought in, it was a brisk sprint through each sale. The hired hands would pull open the gate, ushering in the angry bull. He would spin and stomp around the show pen, twitching his ears and swishing his tail, looking fiercely for an exit. The auctioneer would read the seller's description: "This stylish stout-made bull will be a real herd improver." Then he would call for a starting bid and the first hand would shoot up.

At the front of the bleachers on either side of the entry ramp, the auctioneer's assistants collected the bids. "Four thousand, four thousand, do I hear four?" One of the assistants would yip, and the auctioneer would jump up to the next level. "Four. Who'll give me forty-two five?" Again, a yip— and so on. For the most part, despite the flurry of raised hands and yelps in response, the buyers were conservative, cutting off bidding well below the sale barn average. At one point, the auctioneer, who was working on commission, scolded the buyers for seeming reluctant to go over $5,000. "Y'all are making some great buys on great bulls today. Every other bull sale in the land has been averaging six to seven thousand. Here's a chance to buy some really well-bred bulls at a price you can live with."

Despite the prodding, Rick hung back. If he could get either of the bulls he was waiting on, he said, it would be a good day at the barn. When the first finally came up, Rick shifted in his seat, waiting. The auctioneer's assistant seemed impressed by this bull, too. "I tell you," he said over the loudspeaker. "A yearling weight of plus-one-hundred-and-four, high ribeyes, big scrotal." The bidding started at $3,500, and Rick was quickly in. He was outbid, then came up to $4,000. "Forty-five, forty-five," the auctioneer called. "Who'll give me half?" He got a bid for $4,250 and Rick countered at $4,500. The auctioneer called and called for another bid, before finally closing the sale. "Mark him the buy of the day," the auctioneer said, as his assistant recorded Rick's bid number. With such a low price on that bull, he had enough left in his budget to make a play for the second and soon came away the high bidder on that one, too.

When Rick's second winning bid had been recorded, he made his way to the sale window. He could hardly conceal his grin as he gave his sale tickets to the girl at the counter, her hair up in a bun like a rodeo queen. Rick filled out the check and then went outside to his truck. He backed his trailer up to the loading chute, where two cowhands loaded one bull, prodding it to the front and locking the gate, before bringing in the second. They moved with speed and confidence, leaning their shoulders into the interior gates before throwing the bolts on the locks. With both bulls on the trailer, Rick was quickly in the driver's seat, clearing out of the lot for the next customer to come in.

As we bounced back down Highway 92 toward the grassy acreage north of his house, Rick was chipper and talkative.

"'Mark him the buy of the day,'" Rick said, putting on the auctioneer's tone. He shook his head and laughed. "I don't know about that, but I'm sure pleased."

"The auctioneer's probably never seen a bad bull," I said.

"No," Rick allowed, "but that one's a pretty good-looking son of a gun."

I told Rick that, years ago when I was in Texas making extra money in school by working on a small cattle ranch, I was helping load a bull into a trailer when it kicked the gate back into the face of the ranch hand. The steel bar split his nose from bridge to septum, like a butterflied filet, sending him on a long drive to the emergency room to get stitched back together and leaving me with a lifelong respect and twinge of fear whenever handling bulls. Rick winced at the very idea of it.

"Were they running longhorns?" he asked.

"No, Herefords mostly." I told him that almost all of the longhorns were gone by the time I was down there. You'd see them here and there, but rarely.

"Some people up here were raising them for a while," he said. "They have great, really low-birth-weight calves. So you'd have one longhorn bull to breed with any cows that you were worried about making it through birthing. But they sell like crap at the sale barn, unless you bring in some fifteen-year-old bull with big horns and a nice brindle coloration, something for the Old West nuts who just like the look of them."

"Not so long ago, though, that's what everybody wanted," Rick said thoughtfully, "and here we're back to water. Before everything was irrigated, you wanted a bull that could survive on range with very little water, very little feed."

In the panhandle region of West Texas, where the Texas Longhorn was first created by cross-breeding various European-stock longhorns, the giant rack had been an asset on long cattle drives when red wolves and coyotes would come in to prey on the herds. But when farmers started pumping out of the aquifer and irrigating, they turned to raising short-fiber cotton—and with such remarkable success that they began converting rangeland to cotton fields and herding cattle into feedlots. Built adjacent to cottonseed oil mills, Texas feed yards would fatten cattle on meal and hull byproduct; the hardy but lanky Texas Longhorn was steadily replaced by brawnier shorthorn breeds like the Hereford and Angus, as well as new specialty breeds like the Beefmaster. But the whole system was dependent on pumping groundwater from the Ogallala Aquifer system.

"And down there," Rick said, "it's a perched aquifer. They're mining fossil water." Cut off from the main aquifer system south of the Canadian River, the reservoir recharges so slowly that it would take thousands of years to refill. So as Texas Panhandle farmers cashed in on beef prices and then, in 2005, started chasing big-time commodity profits after passage of the Renewable Fuel Standard ushered in the era of $8 corn, they were pumping out the aquifer faster and faster, draining the great basin of water that had sustained Texas cattle for two centuries—and before long, their wells were starting to run dry.

By the onset of the drought in 2011, the groundwater shortage had grown so severe that the State of Texas commissioned an in-depth study to quantify the problem. The results, published in the Texas Water Report in May 2015, could hardly have been more dire. "Since the 1940s," the study reported, when ranchers first began trying to grow cotton on the arid rangeland, "substantial pumping from the Ogallala has drawn the aquifer down more than 300 feet in some areas." But the real trouble has been much more recent. One hundred feet of the 300-foot decline of the aquifer occurred in the decade between 2001 and 2011. This period coincided with the run-up in commodity prices that tempted farmers to start growing thirsty feed crops. With rising temperatures and what the report described as "the near-total absence of rain" during the drought, water use for irrigation had jumped another 43 percent.

"It's really a cautionary story," Rick said. "And California's in the same fix now, too, pumping their water until there's none left. Nebraska has been smart that way. The NRDs are a model for the country and something to be really proud of. We pump a ton of water from the aquifer, but there's monitor wells and the NRDs tightly watch the levels. If the level drops below the control, where it was when they started monitoring in 1972, they shut everything down. In the summertime, we listen to the radio out of York, and you'll hear days where they say, 'It's a red day; you're on control.' And you don't get more than twelve hours of pumping for three days out of that week."

"How do they make sure you're not irrigating?" I asked.

"They can electronically shut the well down," Rick said, "and you'll hear the old guys complaining when that happens, but it's a good thing. It protects the resource—to make sure it's there when the next generation takes over."

THE SUN was still low on the horizon as Kyle fishtailed down the gravel road. It was early, no one else out yet, and he was just making the rounds, checking on pivots to be sure they were all on and turning. A light rain had fallen overnight, maybe a quarter inch—just enough to leave the morning muggy and windless. But York County was still well short of the four-inch average for the month of July, and it was now into the midsummer run, right before and right after corn pollination, when even the most cautious and judicious farmers turn on their pivots and let them run day and night. The last thing you want during that crucial period is to lose irrigation even for a day.

So far, everything had been fine, but then, just as Kyle was in the homestretch, he spotted a problem. Four miles south of Centennial Hill, in a quarter section planted to corn, a spot the Hammonds call the Metz Place, the irrigation system had stopped moving overnight. Kyle sighed and threw on the brakes. He climbed out and killed the power at the control panel. "The end of July, when the pivots have been running for a solid month, problems start showing up," he said. He grabbed a few tools and his voltmeter from his truck and waded down the rows, making his way toward the tower, its light no longer blinking, at the center of the field.

Fifty years ago, when farmers had to wait until mid-June to plant hand-selected corn seeds, the health of a corn plant was judged by the old saying "Knee-high by the Fourth of July." Today, with hearty hybrids making it possible to plant in early May and selective breeding pushing plants ever taller, corn plants are head-high by Independence Day—and towering taller by the end of the month. Kyle pushed down a dense row, brushing the flat leaves aside as he went, until he finally got to the pivot platform. He climbed partway up the center ladder until he had a vantage above the tassels of corn. "If you look straight down the spans," Kyle said, lining up the arched sections from one tower to the next, "you can see if there's something wrong." About halfway down the line of the irrigation system, one tower was tipped at a slight angle. "You go to the fulcrum of that angle," he said, "and that should be the point of the problem."

Once Kyle got to the tower, though, he couldn't find anything wrong. He checked the contactor and the electrical box. Everything looked right, so he squatted low, studying the wheel mechanism until he saw: one of the wheels seemed to have stuck in the mud or simply seized up from the humidity and the constant use. Either way, when the threaded gear coming off the drive shaft was unable to turn the wheel, the shaft snapped under the pressure. The pivot needed a replacement gearbox, but Kyle wasn't about to call in a Valley repairman. "A new gearbox is probably twelve hundred bucks," he said, "and a service call is eighty-some bucks an hour, and that includes drive time coming from Grand Island." He decided to swap it out himself. "I got a used gearbox in the shed, saved from a pivot that went down in a tornado a few years ago," he said. He was sure he could put on the salvaged gearbox and get the pivot back up and running without losing a precious day of irrigation.

So Kyle went back to the farm and got Meghan. Together, they gathered a handyman jack, a breaker bar, and some giant blocks to lay as a foundation under the jack. "I don't know how long the pivot stuck there," Kyle said, "but with it watering that one spot, plus the rain overnight, the jack will just go right down into the mud." The biggest problem was the sheer heft of the gearbox itself. It must have weighed 150 pounds, and there was no good way to get it from the truck at the edge of the field, halfway across to the central tower. So he parked on the west side, simply hoisted the hulking gearbox onto his shoulders, and started into the field, cutting across the rows.

Kyle kicked the stalks down and stomped them under his bootsole, making a path for Meghan to follow with the jack and blocks, and then he stepped across the furrow and kicked down the next stalk. Losing those plants wasn't ideal, Meghan explained, but better to lose a few ears now than have the whole field go without irrigation for even a day at the peak of the summer heat. "Anything to shorten the trip," she said. Three times as he went, Kyle had to stop and catch his breath, but he didn't quit until he was back at the broken drive shaft. Meghan laid out the blocks, and Kyle jacked up the tower.

By now, the sun was high overhead, beating straight down on the shadeless track made by the pivot's massive tires. The temperature gauge in the truck had been reading over 90 degrees. And, deep in the densely planted field, there wasn't so much as a whiff of wind. "I'm about ready to pass out," Meghan said. "I can't take it." Kyle worked quickly, removing the seized wheel and then unbolting the broken drive shaft and gearbox. But then, when he hoisted the salvaged replacement into place, he had an awful realization. Gearboxes for Valley pivots are mounted in opposite directions, depending on which side of the tower the box is on. The new box matched the old—but as a mirror opposite.

"I grabbed the wrong box," Kyle said.

"What?" Meghan demanded. "Are you fucking kidding me?"

Later, when they were back home and cooled off, Meghan was quick to take her share of the blame. "I could have double-checked with him and made sure he had the right one," she said. "And trying to put an old gear box on the pivot, there was already a risk that it wouldn't work. So I knew going out there that we might have to do it more than once."

"It's definitely not what I wanted to happen," Kyle said, "but at that point, with the whole thing tore down, you know what you've got to do next."

He put the backward gearbox on his shoulders again and started over the path he had made across the rows of corn. They drove back to the shed and got the reverse version, the correct version, of the gearbox—and Meghan grabbed a length of wire. "The old one is too heavy to carry out," she said. "It ain't worth it." They could just wire the broken box to the tower and let it spin around the field from now to harvest time. "Tie it to the damn pivot," she said. "It can hang on there until the corn's out." So they returned to the field and Kyle shouldered the right gearbox out to the tower. He replaced everything, and they wired the broken box to a rung on the ladder that leads up to the electrical box. With everything now in working order, Kyle turned the control panel on again, and the pivot resumed its rotating, the dead gearbox swinging underneath. It had taken all day, until late in the afternoon, to get it fixed, but the pivot was back up and running.

IN SEPTEMBER, Rick kept his promise. When it was time to hitch up the trailer and head west to Curtis, to load up the breeding bulls and bring them back before the fall harvest arrived again, he offered me a spot in the passenger seat. With the empty trailer bouncing and shimmying behind us, Rick didn't want to take the Interstate. Instead, we dropped down to Highway 34 and slowly made our way up the two-lane road. Around Holdrege, the terrain began to soften into sloping hills, and by the time we reached Eustis and Farnam, the loess hills had started to crest and break into blowouts. The deeper we drove into Rick's native terrain, the more he seemed to relax, drawn away for the moment from the worries of prices, the daily fears of soybean and corn futures still sagging ahead of another harvest, but there was one concern he just couldn't push out of his mind. After months of thought—and one final check of the books—he had decided that he had to let Dave, the farm's hired hand, go.

"I thought about waiting until after we got through harvest," Rick said, but Dave had been with him through fourteen harvests. This would be number fifteen. "After that many years with me, I owe it to him to let him make the decision. I think he'll see the year through. Any farmer wants to see the crops that he planted harvested. I hope he does stay through harvest."

We came over a rise, and suddenly the cropland dropped away, giving way to grass, knee-high to hip-tall and yellowed with the coming of fall. Rick looked down the highway, empty ahead of us for as far as the eye could see. "Here's the deal," he said, turning suddenly frank. "When that neighbor took a half section away from us, I knew right then, and that's been a year and four months ago, but I sat on it for a year trying to see what would happen, trying to find some way to keep Dave on. But over time, three things have become apparent. One, we don't have enough work. Without the hogs, big distractions like the new house or building a barn, we can finally concentrate, and we can handle the work. Two, with the collapse in commodity prices, we can't afford it. That's just reality. The prices are way down, and I don't know if they're ever coming back. The third thing is that we have Kyle now—he's like having four hired men—and Jesse is coming back for at least half the year now."

Rick had adopted a tone of finality, but his body language suggested resignation more than resolve. He didn't like having to let Dave go, but he knew it was the right thing for his family and their business. "It's just time," Rick said finally. "It's time for it to get passed down to the next generation. I've been mostly going through the motions for a long time. It's time for me to step aside, for Dave to move on, and let Kyle and Meghan and Jesse take over."

"How will you divide everything?" I asked. Remembering the years of tension among Heidi and her sisters, I couldn't help wondering if Rick was worried that there were similar fights ahead among his own children. He shook his head ruefully.

"The way that the Harringtons did things was an undivided share. I say that is a recipe for strife. It's not going to happen that way with us. Each of the kids will have their own parcels, and they'll make their own decisions on their own ground, and their own marketing." He even thought they should buy their own equipment, rather than try to share the costs on major purchases, like tractors or sprayers. "You see so many of those deals where the families end up in fights. They hate each other. I'm not going to have that." Instead, he was resolved that each kid would have an equal share to farm or sell as they saw fit, and it wouldn't be gradual; it would be fast.

"Meghan and I went to this workshop on succession," Rick said. "They said, 'Do it in chunks, because there's these beginning farmer interest rates.' My original plan was to let them take over rented ground. But the tax adviser said, 'Look, if you give them *your* ground, then there's a huge state income tax benefit.' That would be the logical thing to do. So we'll get them through this year, and then I'll probably give them some of our owned ground. We'll work it out, and maybe we could even do the cattle. Sell it all in one chunk." He and Heidi would keep their dream house in Hordville, and he would hold onto this land in Curtis, where his sisters and brothers-in-law lived. But it was time for him to make way for the kids.

When we finally arrived at the ranch, it was too late in the day to load the bulls and get back before dark. Rick pulled the trailer to the edge of one of his fields and unhitched it, then we drove out in the grass, just to make a quick check on the cattle grazing there. He banged on the door and hollered, and the black cattle came like shadows moving through the grass. They poured over the lip of the buttes and scrambled down the gray trails they had worn. He herded them through a gate into a lower pasture, so he would have an easier time of gathering up the bulls in the morning.

Before dark, Rick drove me by the house where he had lived as a young boy. "This was the center of my life, this barn," he said pointing. "My dad had cut the center of the hay-mow out. We had a big rope all the way to the roof, and we'd swing catty-corner from one end of the barn to the other." He'd play until after dark. By then, when he was supposed to gather eggs from the henhouse to bring in for breakfast in the morning, he was certain the shadows were filled with coyotes or wolves. To stay calm, he'd list off all the cartoon characters he could think of. *Mickey Mouse, Donald Duck*, he'd say to himself, scanning the edge of the trees. *Goofy, Daffy, Min-*

nie. "And when I got as far as the well, I'd just tear off to beat hell," he said, laughing. By then, he was sure that he was close enough to the house to outrace whatever might come clawing out of the dark. I laughed too, but I couldn't help thinking of what Rick had told me before—how he'd spent his whole life just barely keeping the wolves away from the door and how as soon as he felt safe, he'd take another chance again.

As we wheeled up the hill toward the house where Rick had lived through middle school, where his sister and her husband now lived, I wondered what Rick would do now. If there were no more wolves, no more fights left ahead, would he enjoy his retirement, or would he live out his days regretting the mistakes of the past? "You know, Kyle and Meghan coming back to the farm has been just a real joy to me," Rick said, as if he knew exactly what I was thinking. "Now, Jesse's starting to come back, too. Evidently, I wasn't quite as mean as I thought I was when they were growing up—didn't make it quite the hell that it seemed to be to me. I worry about how they'll do, but I think it'll be just like when I started. A light bulb goes off when it's yours. You just think about it differently."

He pulled into the driveway until he reached the yard of the old farmhouse. "You think about it probably too much," he said. "I know I did." Inside, Rick sorted through more old memories. In the basement, he fiddled with the lock on an old chest, until the key finally turned and revealed the few things inside. Stacks of Rick's *Classic Comics* with illustrated versions of *Robinson Crusoe* and *The Count of Monte Cristo*. There were a few photographs, including Rick's high school graduation photo. "Look at that guy," he said. "Have you ever seen somebody with such confidence?" He laughed to himself. "Yeah,

I had it all figured out back then. I keep getting dumber the longer I live."

Finally, Rick closed everything up and stepped back into the evening light of the yard. He wandered down the slope through the yellow grass. He found a turkey feather and stuck it into the band of his cowboy hat. For a moment, I thought he might make another joke about that, but he had turned thoughtful. The windmill squawked quietly behind him. "C'mon," he said. "There's one last thing I want to show you." He walked down until we were almost at the creek bottom. There, among the trees and overgrown by vines and saplings, was the ruin of a tiny bunkhouse.

"This is where our farmhand lived," Rick said. He ducked through the doorway, nothing but rusting hinges guarding its entrance now. "His bunk was right there. This was his stove. As I got older, I loved to come down here, just to listen to the way grown men talked," he said. "He would roll his own cigarettes. I remember how he stopped in the middle of his stories while he pinched out the tobacco and licked the paper."

The windows of the bunkhouse were all cracked and broken out, the little glass left behind filmed over by ancient grime. "My dad had to let him go when things got bad," Rick said. "I don't know what ever became of him."

THE TRI-STATE

BAYARD, NEBRASKA

September 2015

I pulled the truck into the shade of a tiny clutch of cottonwoods. After the trip out to Curtis with Rick, I couldn't stop wondering: what makes a homeplace? For Rick, it seemed to me, Curtis was where his heart had remained and would always be. It's where, he told me, he wanted his ashes scattered. As a kid, I had felt a similar tug, drawn to the tiny town where my dad grew up, so I talked him into making the drive there, seven hours across the state. The town, Bayard, had been slowly disappearing from the time I was a kid, visiting my grandparents over the summer. And with every disappearance—a restaurant closing, a shop boarded over—there was an equal disappearance from the surrounding neighborhoods and farmhouses, another family that had decided that this was no longer home for them.

By the time we arrived, the sun was sinking in the west,

but the temperature was still somewhere in the 90s, the song of the cicadas ratcheting up in the distance. I followed my dad out onto the sandy path that connects the state highway to the Tri-State ditch road north of town. A tiny house with a fenced patch of dirt sat just back from the canal; during the early 1950s, when Dad was still in elementary school, he lived there with his parents. "It looked just about exactly like this back then, except for the fence," Dad said. "That would have saved me a lot of grief." As it was, he said, he endured years of parental warnings: never to venture too close to the ditch bank, never ever to cross "the needles"—a concrete foot-bridge and series of weatherworn pickets that form a weir across the channel.

Dad explained that there is a narrow lip at the bottom, just on the upstream side of the bridge's pilings. Ditch riders thread beams into the current, searching for that foothold, then let the force of the water push the top of the pole tight into place. By effectively narrowing the width of the canal, the needles form a bottleneck that partially dams the current and raises the flow level during dry times, but they also cre-ate a rushing spillway on the backside—a constant source of worry for my grandma. Maybe she fretted more than most mothers, having lost two children in infancy before my father was born, but her fears were far from unfounded. During the years Dad's family lived along the Tri-State, two young children, in separate incidents, drowned in sections of ditch that ran through nearby Scottsbluff. "There were kids that drowned, no doubt about it," Dad said. "And it was cer-tainly a worry."

But for my grandparents, there was little choice. My grand-father had left the Omaha stockyards to come west in 1936, then worked as a tenant farmer, raising beets and beans along the Platte River bottoms south of town. My grandma got them through lean seasons on her teacher's salary in the proliferating county schools. They hung on, year to year, until the blizzard of '49 pushed the family over into bankruptcy. Farmers Irrigation District, which still owns the Tri-State Canal, was willing to provide a rent-free home and an old crank telephone (the first my dad's family ever had), so that ditch riders could be called to go out and adjust water levels immediately in cases of emergency, and ditch riders, in turn, could report washed-out sections or other problems back to the offices in Scottsbluff. Ditch riding was seasonal work back then, but they could have the house year-round, and my grandfather got work in the winters working as an oiler in the pulp drier at the Great Western Sugar factory in town.

During the summers, when school was out, my dad would ride along as Grandpa went on his rounds, checking to make sure farmers along his five-mile section of ditch were getting their allotments. "The ditch riders had to supply their own cars," Dad said. It was a black coupe, he remembered, but paused a moment trying to recall the model. Water crawled through the channel, then rushed through the needles. "A Chevrolet of some sort," he said finally. "He always bought Chevrolets; he would never have dreamed of owning a Ford. Mostly what I remember is that I could barely see out." At each farm, Grandpa jotted into a pocket notebook how many acre-feet of water each farmer was receiving that day, based on how much water was being released out of the Guernsey Dam in Wyoming. At the end of the line, he would take a final reading to make sure the Tri-State was delivering the required amount into the Northport Canal.

"I remember one time, down real near to the Northport check, out in that sandy soil, he could see water bubbling up and obviously coming out of the ditch. He tried to plug it at the end out in the field, but he couldn't do it," Dad said. "So he had to get down into the ditch and feel around for that hole. He took a bunch of gunnysacks and jammed them in there, then packed some mud in to try to cut off the flow of water. If you didn't, eventually it would cut the ditch bank wide open, and the water would run out and flood everybody."

But most days, he was just checking head gates, to make sure that there was the right amount of water in the lateral ditches and flowing into the farmers' open irrigation channels. "There were always farmers who were upset that they didn't get water when they wanted it or they didn't get enough," Dad said. "And they weren't above trying to steal a little water from their neighbors or taking a little extra out of the ditch." Grandpa understood that they were just trying to keep from going broke, but he also knew, Dad said, "you're entitled to a certain amount of water, and that's what you're going to get. It's the ditch rider's job to see that you get all of that, but he also sees you don't get any more so that you short somebody else down the line."

Dad pointed toward an invisible point in the east. "Remember that big lateral we saw down there," he asked, "the one at the foot of that hill where the feedlot is now?" I nodded. He said he remembered one instance when Grandpa suspected the farmer who used to live there of stealing water. Every morning Grandpa would go on his morning ride, checking the headgates. And then again, each afternoon, he would ride back along the same route, adjusting as needed. At that property, he started noticing that the ditches seemed overfilled each afternoon and began to suspect that the farmer was coming along behind him, opening the gates wider each morning and then cranking them back down before Grandpa made his afternoon rounds. "So one day," Dad said, "instead of going home and calling the readings in, he went back to that headgate. And, sure enough, it was wide open." So Grandpa went to his trunk, took out a big log chain, and wrapped it around the crank of the headgate and locked it—*open*. "So when the farmer came back to close the gate back down that afternoon," Dad said, "he couldn't turn the crank." He was forced to make a frantic call to Grandpa, confessing to stealing the water, and begging him to come down and unlock the gate before it washed out all of his ditches and flooded his fields.

"This is beautiful country," Dad said, as we shadowed the ditch back toward Bayard. "It was a great place to grow up, but I couldn't get away from here fast enough." He said he didn't want to spend his life at the mercy of banks, insurance adjustors, and every storm cloud forming on the horizon— every day a fresh (but identical) struggle to beat back total loss. I understand it, too, of course. Raised in the city, with only vacation trips to Scottsbluff and Bayard, it was easy to be nostalgic for a vanishing way of life when I didn't ever have to suffer through its hardships. Still, passing back through town, past all of the shuttered storefronts and abandoned buildings, I lamented aloud how sad it was to see Bayard dying this slow death.

"Is it?" Dad asked. He sounded somewhere between quizzical and professorial. "Maybe this town has served its purpose, and now it's time for it to fade into history."

I turned down toward the old Great Western Sugar factory—shuttered after the end of the last sugar campaign that fall—and was surprised to find it not only abandoned but unfenced and unguarded. I drove us around the plant to the dryer where Grandpa worked. The giant doors stood open, so we parked and walked in. All that colossal machinery (the coal scuttle, the enormous furnace, the blower the size of a semi) bore mute testimony to just how dominant the beet industry had once been. And how utterly—rusted now and forgotten—it has been overtaken by a monoculture of corn. The warehouse at the back of the building is big enough to house a football practice facility, but the wind whistled and made its corrugated siding creak, sending unnerving echoes bouncing through its empty rafters. It was hard not to see this as a symbol of the end of the era of community and shared purpose that founded the valley.

Today most farmers and ranchers are forced to go for broke each year, planting fields of whatever crops and grazing herds of whatever breeds are selling highest on the futures market. They bank on their ability to work harder and raise more than their neighbors, on the prospect that enough of them will fail that prices will remain high at harvest time. Now that ethic comes to a crossroads, as everyone faces the real possibility that a resource as basic as water—worried over and doled out and preciously conserved for more than a century—may be in diminishingly short supply. But no one seemed to be thinking beyond next year. No one would discuss the possibility that this drought was more than a few hard years, that there might be many hard years brought on by climate change and rising temperatures. For all the careful stewardship of the land, few in the North Platte River Valley want to consider that they may be up against a new threat, bigger and more complex than the one that faced farmers and ranchers at the end of the nineteenth century—one that requires them to think even more boldly than their ancestors. For now, they're resigned to waiting and praying for snow in the Rockies.

Before we headed back east, my dad and I stopped at Genoways Hall, the public meeting room built by the Civilian Conservation Corps and named for my grandfather even before he died. The Bayard ditch runs within feet of its side door. It's not part of the Tri-State Canal system where he worked, but I remember how, each time we went there (when I was still too young for school), he would stalk the ditch bank, breaking back weeds and clearing debris. And I remember taking my father's old wind-up motor boat, filched from his wooden toy box on the porch of my grandparents' house, cranking its key tight, and then letting it go. Most of all, I remember how it disappeared into the pipe that still runs under Main Street, then popped out on the other side, listing drunkenly in the current; how I held my breath—afraid of something, I'm still not quite sure what—until my Grandpa patiently fished it out of the reeds and motioned with a wave of his sun-browned hand that it was time for us to get back home.

WELCOME NEWS

November 2015

On November 6, 2015, Meghan and Kyle were in Muir Woods in northern California at their wedding rehearsal. They had decided on a destination wedding with just family. In the middle of the rehearsal, Meghan's phone started to blow up. Texts would come through asking if she had heard the news, but she couldn't get a phone signal to call out. Finally, at the top of the ridge, overlooking the tops of the redwood forest and out on the Pacific Ocean, she was able to get reception on her phone.

President Obama had rejected the Keystone XL pipeline, putting an end, at least for the moment, to six years of fighting and worrying about the fate of the Ogallala Aquifer if there were ever to be a spill. Secretary of State John Kerry had decided that the project was not in the country's national interest, and Obama had called a press conference at the

White House to announce that he supported that opinion. "America is now a global leader when it comes to taking serious action to fight climate change, and frankly, approving this project would have undercut that leadership," Obama said.

Meghan hooted and screamed and started calling everyone back home in Nebraska, all the people who had been fighting at her side for years. "I can't believe it, I can't believe it," she said. "It's the best wedding present I could ever ask for." The next day, Meghan and Kyle were married, and when they came back to York County, they soon moved into the house that Dave and his family had vacated, the home where Meghan had spent her childhood on the northern edge of Centennial Hill Farm. Less than a year later, Meghan found out that she was pregnant.

Despite all of the happiness, the future remains uncertain. In 2014, Rick had warned that the American farm might be in trouble if we saw two more years of record harvests. Now, that's exactly what has happened. The harvest of 2016, another record-setting year of production of core commodity grains like corn, soybeans, and wheat, pushed prices to their lowest levels in decades. Corn, in particular, has plummeted to less than half its market value of just five years ago. Despite the drop in prices of feed grains, livestock prices have fallen simultaneously and at a rate that made it impossible to capture profits on increased margins. This downward spiral is already having broader effects. Cash-poor farmers aren't updating equipment or buying new trucks or even going to town to spend money on food and entertainment. The rural economy is stalling.

Worse still, farmers who took out loans for land or equipment at the peak of prices are starting to worry about their ability to service their debts with profits projected at the time—and banks are growing nervous, too. Loaning institutions are starting to call on big farmers to liquidate landholdings used as collateral, in order to reduce their risk. But this trend is already dragging down property prices, forcing still more liquidation—the exact cycle that led to a rapid devaluation in the 1980s, triggering the Farm Crisis. Today the potential dangers of a rural bank panic to the broader economy are even greater. Three-quarters of farmers have quit the business in the last thirty years, so now every failure carries four times the weight.

To break this downtrend, the American Farm Bureau Federation had been counting on President Obama's Trans-Pacific Partnership, which promised to open new markets for beef, pork, and grains. But the rural areas that were depending on this new deal, as well as standing agreements such as NAFTA, voted overwhelmingly for Donald Trump and his protectionist, antiglobalist policies. Now, Trump is threatening to cancel manufacturing trade deals with China, and China is responding by threatening to cancel its purchases of American grains. If such a thing were ever to happen, it would make the Farm Crisis seem like a minor economic ripple.

Still, Meghan and Kyle have taken over their share of the family land, and in February 2017, they welcomed the seventh generation—a girl, Logan Elena Galloway— to Centennial Hill Farm. "I don't know where we'll fit into the grand scheme of things," Meghan told me one day in the shed at the farm. "I just know that we want to do the best that we can for the land, for our family, for the world. Whatever comes, we'll find a way to get by."

Acknowledgments

First and foremost, I have to thank Rick Hammond, Meghan Hammond, and Kyle Galloway for allowing me to follow them for over a year of their lives. I was constantly in the way, parking in the wrong place, asking complex questions at the wrong time, but they handled it all with patience and grace. Thanks, too, to Jenni Harrington for helping me to sort out the history of the Centennial Hill Farm and to Rita Zoucha for generously allowing me into her home and sharing information with me about her son Brent. Thanks to my father, Hugh H. Genoways, for answering questions about our family history. And my gratitude to the countless people, unnamed and unquoted, who contributed to my understanding of modern farm life. This city kid appreciates the education.

Second, I would like to thank the many editors who believed that this story was worthy of telling. Most importantly, John Glusman at Norton saw a book here and helped in countless ways to make this project into one. Everyone at Norton was so supportive and engaged: Lydia Brents, Fred Wiemer, Alexa Pugh, and Will Scarlett. Thanks also to: Ted Ross at *The New Republic*, who was brave enough to publish a long piece on the history of corn hybridization; Ellen Rosenbush at *Harper's*, who shepherded the story of a soybean harvest through trying times; Jennifer Sahn at *Pacific Standard*, who made immeasurable contributions to shaping the section on center-pivot irrigation. The Food & Environment Reporting Network (FERN) provided much-needed funding to keep this project going; special thanks to Sam Fromartz, Brent Cunningham, and Tom Laskawy. Thanks to Don Fehr and everyone at Trident Media for standing behind this project and everything I do.

Last but not least, my endless gratitude to my family. My son, Jack, lived through nearly three years of trips to the farm and conversation about market prices, equipment failures, and water rights. My wife, Mary Anne Andrei, made the first contact with the Hammonds, conceived of a long project following them, and was frequently the glue that held it all together. The intimacy of her photographs is a testament to how much the family trusted her and an indication of the extent to which my access was granted because of her legwork. I'm grateful for every day that we're allowed to work together, telling stories that matter to us.

Selected Bibliography

The majority of my research came from observation and conversations with the Hammond-Harrington family, as well as other sources directly quoted in the text. However, a few books were critical to shaping my thinking and informing the text.

Deborah J. Bathke et al. *Understanding and Assessing Climate Change: Implications for Nebraska*. Lincoln: University of Nebraska Press, 2014.

Ann Bleed and Christina Hoffman Babbitt. *Nebraska's Natural Resources Districts: An Assessment of a Large-Scale Locally Controlled Water Governance Framework*. Robert B. Daugherty Water for Food Institute. University of Nebraska, 2015.

Wayne G. Broehl, Jr. *Cargill: Trading the World's Grain*. Hanover, NH: Dartmouth College: University Press of New England, 1992.

Dan Charles. *Lords of the Harvest: Biotech, Big Money, and the Future of Food*. New York: Basic Books, 2001.

John C. Culver and John Hyde. *American Dreamer: A Life of Henry A. Wallace*. New York: W. W. Norton, 2001.

Dan Morgan. *Merchants of Grain: The Power and Profits of the Five Giant Companies at the Center of the World's Food Supply*. New York: Viking, 1979.

Ronald C. Naugle, John J. Montag, and James C. Olson. *History of Nebraska*. 4th ed. Lincoln: University of Nebraska Press, 2015.

Quentin R. Skrabec, Jr. *The Green Vision of Henry Ford and George Washington Carver: Two Collaborators in the Cause of Clean Industry*. Jefferson, NC: McFarland & Co., 2013.

Nancy Warner and David Stark. *This Place, These People: Life and Shadow on the Great Plains*. New York: Columbia University Press, 2013.